W9-DEA-100

WITHDRAWN

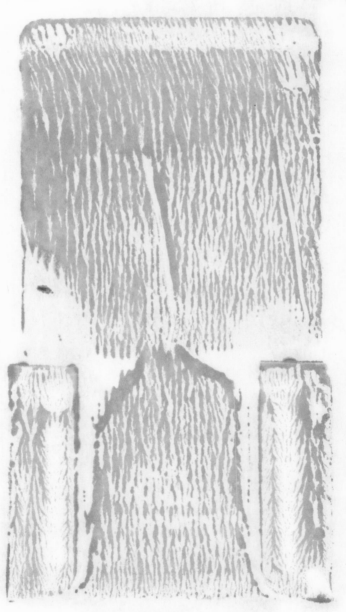

Nuclear Materials

NUCLEAR ENGINEERING FUNDAMENTALS

These five books were written for the wide spectrum of people interested in nuclear physics and engineering who are not themselves physicists or nuclear engineers, but who share a common need for an introduction to nuclear concepts.

Nuclear Materials

ALVIN BOLTAX, Sc.D.

Westinghouse Electric Corporation
Astronuclear Laboratory

EDITED BY
ROY WEINSTEIN

McGraw-Hill Book Company

New York   Toronto   London

NUCLEAR MATERIALS

Copyright © 1964 by McGraw-Hill, Inc. All Rights Reserved.
Printed in the United States of America. This book, or parts
thereof, may not be reproduced in any form without permission of
the publishers.

Library of Congress Catalog Card Number: 63-11856

69043

PREFACE

The problem of materials for use in the construction of nuclear reactors is highly technical. It requires a thorough understanding of nuclear physics and metallurgy and a broad background in materials engineering. For example, any problem involving the use of materials for particular applications will necessarily involve the physical and chemical properties of materials, processing and fabricating techniques, mechanical strength, corrosion resistance, etc. With these facts in mind, Book IV deals with the basic metallurgical problems faced in choosing materials for nuclear reactors. As an introduction to metallurgy, and in particular to the physics of metals, Chap. 3 attempts to answer the always perplexing question: What is a metal? Chapter 4 contains a survey of the broad field of metallurgy starting with ore processing and ending with fabrication techniques such as casting, working, and welding. Chapter 5 explores some detailed reactor-materials problems such as radiation damage, corrosion, and thermal problems. The remaining chapters deal with such problems as functions of reactor components and detailed information about reactor fuels, constructional materials, and reflectors and moderators.

The author wishes to thank Dr. Robert O. Teeg for proofreading most of the manuscript. Dr. Teeg also collaborated in writing some of the material in Chap. 7. Thanks are also extended to Mary Ferreira and Barbara Rogen for typing the manuscript.

ALVIN BOLTAX

CONTENTS

Nuclear Materials

REACTOR MATERIALS

1·1 Introduction

From the beginning of the atomic energy program, scientists and engineers have been faced with the problems afforded by completely new materials. To the usual design considerations of physical and chemical properties has been added a completely new materials criterion, nuclear properties. The materials acceptable for reactor use have been found mainly among unusual materials and among some exceedingly pure common ones. Extensive development programs have been carried out by the U.S. Atomic Energy Commission (USAEC) to devise methods for producing commercial quantities of exceptionally pure elements and compounds from materials many of which previously had been laboratory curiosities.

The difficulties encountered at the beginning of this program were staggering. Until 1940, for example, the total production of uranium in this country was not more than a few grams, and even this was of doubtful purity. At the same time, the nation's output of metallic beryllium and zirconium would barely have filled a matchbox. Carbon, necessary as a moderator material, had never been produced in quantity with anywhere near the purity required. In addition, there were huge voids in the knowledge of nuclear constants, and no one was able to predict with confidence how various materials would behave under pile irradiation.

Since 1940, there have been remarkable accomplishments in the development of materials for atomic weapons and low-temperature research reactors. In this connection a low-temperature reactor would operate at temperatures below 400°C. The problem of reactor materials is much more acute in the case of high-temperature power and breeder reactors. The ordinary high-temperature engineering materials are alloys containing cobalt for which the neutron-capture cross section is much too high to permit their extensive use in thermal reactors. High-strength nonmetallics such as refractory oxides, carbides, and nitrides cannot be used generally because they suffer from poor resistance to thermal shock (stability under conditions of rapid thermal variations).

The common low-cross-section metals, aluminum and magnesium, are definitely not suitable at high temperatures, because of their low melting points. Only recently have zirconium and beryllium been produced in pure form in commercial quantities, and even these metals cannot be used un-alloyed at very high temperatures. Thus, the selection of reactor materials has become a continuous struggle to overcome the limitations of common materials and to discover new and more suitable ones.

This short introduction gives an idea of the materials problems facing nuclear engineers. The approach to a partial solution of some of these problems, which will follow, will be carried out by a threefold attack. First, the functions of the various materials which comprise a nuclear reactor will be discussed. The second step will involve a detailed discussion of the scientific techniques used by nuclear engineers to evaluate the properties of materials. Finally, detailed information about the properties of the most widely used reactor materials will be examined and tabulated.

1·2 Comparison of a combustion furnace and nuclear furnace

Fundamentally, a nuclear reactor used for creating power is a furnace. The fuel in the case of a nuclear reactor can be one of several fissionable materials such as U^{235}, U^{233}, or Pu^{239}. Conventional furnaces use mainly coal or oil as fuel materials. The reaction liberating energy in our modern-day furnace is basically oxidation. This is a reaction which results in the combination of fuel atoms with oxygen atoms. The symbolic representation for this reaction is given in Eq. (1·1).

$$F_{\text{fuel}} + O \rightarrow FO_{\text{compound}} \tag{1·1}$$

To give an example of the energy liberated in such a reaction, if 1 kgm (2.2 lb) of pure carbon were completely oxidized to carbon dioxide (CO_2), about 8000 cal would be released.

In a nuclear reactor the energy-liberating reaction is fission, i.e., the splitting up of an unstable nucleus into two smaller nuclei. For U^{235}, the reaction might be represented by Eq. (1·2).

$$_{92}U^{235} + _0n^1 \rightarrow _{57}La^{147} + _{35}Br^{87} + 2_0n^1 \tag{1·2}$$

If 1 kgm of U^{235} were fissioned in the manner described above, approximately 20×10^{12} cal would be released. The energy released by fission of 1 kgm of U^{235} is over 2 billion times that released by complete combustion of 1 kgm of carbon.

We continue the comparison between a furnace and a nuclear reactor by examining other components of these types of power sources. The reaction chamber of a furnace is completely surrounded by insulating material and thermal radiation shields. In an analogous manner the nuclear-reaction chamber is surrounded by reflectors, moderators, and radiation shields. In both cases, the surrounding material is designed to prevent energy losses from the reaction chamber. The nature of the insulating material for nuclear reactors will be discussed further in the following chapter.

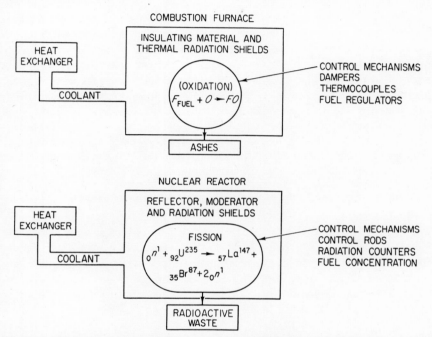

Fig. 1·1 Schematic comparison between a combustion furnace and a nuclear reactor.

The control mechanisms for the two power sources must allow maintenance of a desired reaction rate. A combustion furnace is usually controlled by the amount of oxygen supplied to the reaction chamber. A nuclear-fission furnace is controlled by establishing a desired neutron leakage or nonfissionable neutron capture by means of control rods. In this case, the oxygen and neutron supply are analogous in terms of the two different furnaces. Finally, the structural materials and heat exchangers for each of the two furnaces are identical in principle. Small differences are encountered in these functions except where corrosion and nuclear constants are involved. For example, the high flame temperatures in combustion furnaces place stringent restrictions on the corrosion properties of materials used near the flame.

Figure 1·1 summarizes most of the above discussion. It is hoped that this analogy will serve to orient the student with reference to the basic components of a nuclear reactor. The next chapter will be devoted to explaining the detailed functions of each of the reactor components, which include fuel elements, moderators, reflectors, control materials, coolants, structural materials, and shielding materials.

2

FUNCTIONS OF NUCLEAR-REACTOR COMPONENTS

2·1 Introduction

A nuclear reactor is a highly complex unit consisting of essentially seven different components. These components are fuel, moderator, reflector, control, coolant, structure, and shielding. Each of the seven components has particular functions to perform for the satisfactory operation of the reactor. Many of the nuclear functions, such as moderating ratio and neutron scattering properties, have been described in great detail in the previous chapters. In this chapter we shall examine some of the physical functions of reactor components, such as strength, corrosion resistance, and thermal properties.

As a supplement to the discussion of the functions of reactor components, we include a survey of the materials possessing the required properties. It is hoped that the student will thus learn to associate certain properties and functions with specific materials. As an aid to the student, Table 2·1 contains a list of materials generally used in reactor construction and a brief description of some of their important properties.

2·2 Fuel elements

Reactor fuel elements usually are made up of a core, consisting of fissionable material dispersed in a diluent, and a cladding to enclose the core and protect it from corrosion and to protect the coolant from fission-product contamination. In converter- or breeder-type reactors the fertile material (either U^{238} or Th^{232}) also may be mixed or associated directly with the fuel. The fuel-element assemblies must meet the requirements listed below:

1. They must have adequate strength under the most adverse conditions of temperature, irradiation, external loading, and weakening caused by burnup.

2. They must resist corrosion by neighboring materials, coolants, and atmosphere.

3. They must be stable, both dimensionally and with respect to mechanical properties, under the operating conditions.

4. They must possess good heat-transfer properties, such as high thermal conductivity, large surface area, thin sections, and no "dead" spots (sections of the fuel element which are poorly cooled because of low coolant flow).

5. They must have good nuclear characteristics, such as low parasitic-capture cross section and low poison-forming tendencies.

6. They must be amenable to ready fabrication, installation, removal, and reprocessing.

The selection of specific materials and designs for fuel elements depends on such general considerations as whether the reactor is to produce power

Table 2·1   Materials used in reactor construction

Uranium Plutonium Thorium	Fissionable and fertile elements. Generally used as an alloy, ceramic (oxide or carbide), or cermet (see below)
Aluminum Magnesium Zirconium Beryllium	Metals with low capture cross section for thermal neutrons. They can be used as cladding materials for thermal reactors. The choice is decided by temperature of operation, corrosive environment, and other factors. Beryllium is also used as a moderator
Niobium Molybdenum Tantalum Vanadium Tungsten	Refractory metals for use in fast reactors where their capture cross sections are acceptable. The choice at present is influenced by many factors, which are covered in the text. Niobium is also a possible metal for use in thermal reactors
Sodium Sodium-potassium alloy Lithium Bismuth Cesium	Liquid metals for use as heat-transfer media. Lithium-7 and cesium have attractive properties. Bismuth has been considered for use as a solvent for uranium in a liquid-metal-fuel reactor
Constructional steels	These range from mild steels (low alloy) to fully austenitic steels and are usually chosen on the basis of their weldability, resistance to particular corrosive conditions, high temperature strength, cost, and other factors in specific cases
Cermets (combination of ceramics and metals)	These will normally be used as fissionable or fertile materials or as a mixture of fissionable, fertile, and moderating elements

alone or breed fissionable material; whether it is to be thermal, intermediate, or fast; availability and economics of reactor materials, temperature of the reactor core, corrosiveness of the coolant; and many other factors.

Table 2·2   Data on reactor fuel-element materials

Reactor type	Reactor designation	Fuel	Cladding
Pressurized-water	Shippingport	Enriched uranium-zirconium alloy	Zircaloy-2
	Yankee	Enriched UO_2	Stainless steel 348
	Indian Point	Enriched ThO_2-UO_2	Stainless steel 304
	Savannah River	Natural uranium	Zircaloy-2
Boiling-water	Dresden	Enriched UO_2	Zircaloy-2
	Elk River	Enriched ThO_2-UO_2	Stainless steel 304
	N. States	Enriched UO_2	Al–1 % Ni alloy
Organic-moderated......	OMRE	Enriched UO_2 dispersed in stainless steel	Stainless steel
	OMRE-Piqua	Enriched uranium-molybdenum alloy	Aluminum
Sodium-cooled	SRE	Enriched uranium	Stainless steel 304
	Hallam I	Enriched uranium-molybdenum alloy	Stainless steel 304
	Hallam II	Enriched UC	Stainless steel 304
Fast-breeder	EBR-I	Enriched uranium-zirconium alloys	Zircaloy-2
	EBR-II	Enriched uranium-fissium alloy*	Stainless steel 304
	Fermi	Enriched uranium-molybdenum	Zirconium
Gas-cooled	Bradwell	Natural uranium	Magnesium alloy
	AGR	Enriched UO_2	Beryllium
	EGCR	Enriched UO_2	Stainless steel 304

* Uranium-fissium alloy is a complex alloy containing the approximate distribution of elements which would be produced by the normal fission reaction. The alloying elements generally include Mo, Ru, and Zr.

Table 2·2 gives a summary of the fuel-element materials for several nuclear reactors. The fuel elements may be divided roughly into three types: slugs, clad plates of "sandwiches," and liquids or loose powders. Figure 2·1 illustrates the general characteristics of fuel slugs and clad plates. Existing reactors using natural or slightly enriched uranium generally employ slug-type fuel elements. Some of the more recent solid-fuel reactor designs replace the canned uranium slugs by clad plates having a uranium alloy or cermet core and a bonded coating of some suitable alloy such as aluminum-tin, zirconium-tin, or stainless steel. Aluminum-alloy-clad fuel elements can be used only for low-temperature applications [up to about 400°F (205°C) with water] because of their poor high-temperature strength and corrosion resistance. Zirconium and its alloys are somewhat better, although their

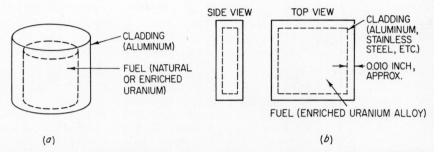

Fig. 2·1   General characteristics of fuel slugs and clad plates. (a) Fuel slug; (b) clad plates, or "sandwiches."

upper limit is probably under 1000°F (538°C). Certain of the stainless steels have very good strength and corrosion resistance, and molybdenum is promising up to temperatures as high as 2500°F (1370°C).

2·3   Moderator materials

Thermal reactors require moderating materials which should be capable of reducing neutron energy rapidly. As discussed in the previous books, materials of low atomic weight are necessary. These materials must have a high scattering cross section, a large average logarithmic neutron energy loss per collision, and a low absorption cross section for thermal neutrons.

On the basis of slowing-down power (SDP) and moderating ratio (MR), which are defined, respectively, as

$$\text{SDP} = N\sigma_s\xi = \Sigma_s\,\xi$$

and

$$\text{MR} = \frac{\Sigma_s\,\xi}{\Sigma_a}$$

[see Book III, Eq. (3·23)], it is found that good moderators are composed of such elements as hydrogen, deuterium, oxygen, carbon, and beryllium. Table 2·3 lists the slowing-down power and moderating ratio for the most common moderating materials. These suitable moderating elements may be compounded or alloyed with others, provided that the diluent materials have low capture cross sections and allow a large atomic density of the moderators. Gases are too low in atomic density to be good moderators. Besides the moderators listed in Table 2·3, hydrides, hydrocarbons, deuterocarbons, and other organic compounds are possible moderators, although the usefulness of the organic compounds is doubtful in view of their radiation sensitivity. The important problem of radiation effects on materials will be discussed in Chap. 5.

Beryllium and carbon or their compounds and metal hydrides are the only good, solid moderators. Although beryllium is fairly expensive, difficult to fabricate, and not very abundant in its pure metallic state, its very low capture cross section and high moderating ratio have led to its use in a number of reactors. Its compounds beryllium carbide and beryllium oxide are refractory materials which make good moderators for power-producing reactors that must operate at elevated temperatures.

Table 2·3   Slowing-down properties of moderators

Moderator	Slowing-down power, cm^{-1}	Moderating ratio
Water (H_2O)	1.53	72
Heavy water (D_2O)............	0.370	12,000
Helium*	0.000016	83
Beryllium	0.176	159
Carbon	0.064	170
Zirconium hydride ($ZrH_{1.9}$)......	0.8	56

* At atmospheric pressure and temperature.

Graphite, a form of carbon, was the first reactor moderating material ever used and is the moderator employed in many of the existing reactors. The reasons for its use are relative low cost, abundance, ease of fabrication, and good physical properties. Graphite is a refractory material and can be utilized in high-temperature reactors in a nonoxidizing atmosphere.

Metal hydrides, such as zirconium hydride and yttrium hydride, permit the use of hydrogen as a moderating material at relatively high temperatures (up to 2000°F for yttrium hydride). The principal limitations to the use of these materials in current reactor designs are due to insufficient information available on the physical, mechanical, and chemical properties and radiation stability.

2·4   Reflector materials

The function of a reflector is to scatter, or reflect, as many leakage neutrons as possible back into the reactor. Therefore, the reflector must meet the same nuclear requirements as the moderator: good scattering characteristics and a low probability of neutron capture. Hence, the same materials which are suitable as moderators are suitable as moderating reflectors. Since the primary function of a reflector is scattering, many additional materials which are not considered good thermal-reactor moderators but which have macroscopic scattering cross sections for fast neutrons (0.1 to 4 mev) are usable as fast reflector materials. Actually, there is some advantage in using reflector materials with high inelastic scattering and poor slowing-down characteristics, since parasitic-capture cross sections generally increase sharply as the neutron energy decreases.

Suitable fast reflector elements include those with high atomic density (e.g., iron); those in the upper end of the periodic table for which atomic scattering cross sections are high (bismuth, lead); and fertile breeder materials, for which fast scattering cross sections are also high (uranium, thorium). It should be noted that these elements have generally high densities and that their use as primary reflector constituents reduces the gamma shielding requirements. However, the moderating ability of the shields used with these types of reflectors would then have to be increased to compensate for fast neutron leakage from the nonmoderating reflector.

In addition to the nuclear requirements described above, both the reflector and the moderator must have all the attributes of good structural materials. These include adequate strength, fabricability, thermal and radiation stability, and corrosion resistance. Furthermore, they must have good heat-transfer properties to dissipate the energy released by neutrons during attenuation.

2·5   Control materials

The adjustment of neutron flux, or power level, in a reactor can be controlled by altering the neutron leakage, the amount of fuel present, or the neutron losses by parasitic capture. Leakage control may be carried out by changing the reflector, the surface-to-volume ratio, or the amount of moderator. Fuel control involves adding or removing fuel in small integral amounts for proper regulation. However, thermal reactors and low-energy intermediate reactors generally are most easily controlled by changing the amount of absorber present.

Absorbing controls, which may be used in reactors in a number of ways, should have the following properties:
1. A high cross section for the absorption of neutrons
2. Adequate strength for solid rods
3. Low mass, to permit rapid movement with slight inertia effects
4. Good resistance to corrosion by the coolant
5. Stability, both chemical and dimensional, under heat and radiation
6. Reasonable cost (good availability and fabricability)
7. Good heat-transfer properties for adequate cooling

Boron and cadmium have been used most commonly as control materials. Of all the natural elements having high absorption cross sections for thermal neutrons, they more or less meet all the requirements outlined, and their metallurgy is well established. Recently, hafnium, removed as an impurity from zirconium metal, has become available in significant quantities.

Gold, rhodium, and iridium, while moderately attractive from cross-section considerations, are not particularly suitable as the principal constituents of control rods. Their structural properties are poor, and their mass densities are rather high for rapid movement with low inertia. In addition, they all form highly radioactive isotopes upon absorbing neutrons.

The low melting points and structural weaknesses of iridium [melting point of 156°C (313°F)] and mercury [melting point of −39°C (−38°F)] prevent their use as solid control materials. The remaining elements of high absorption cross section are rare earths, which are not available in commercial quantities at present. Intensive USAEC-sponsored research since the end of World War II has greatly increased our knowledge of rare earths, and it is possible that some of these materials would be on the market today, at a reasonable cost, if the demand existed. Some of the heavier rare earths show characteristics which might allow their use as absorbing controls in epithermal or intermediate reactors. The capture cross sections of these elements do not decrease as rapidly with increasing neutron energy as they do in most materials.

There are a number of elements which have isotopes with unusually high cross sections. Examples of these are B^{10}, Cd^{113}, Sm^{149}, Gd^{157}, Yb^{168}, and Hg^{196}. Such isotopes might be used for control purposes if they were easily obtainable, perhaps as a by-product of some other separation program or as fission products. One point which requires further consideration before these isotopes can be used is the nature of the product resulting from neutron capture. Cd^{113}, for example, transforms by an (n,a) process to stable Cd^{114}, so that the strength and corrosion resistance of a cadmium control rod containing a large percentage of Cd^{113} would be virtually unaffected by high burnup. On the other hand, B^{10} transmutes to stable Li^7 by an (n,a) reaction, with resulting undesirable change in properties as the B^{10} atoms are used up.

2·6   Reactor coolants

The nuclear requirements for the medium that removes fission heat from the reactor are similar to those for the moderator, because of the relatively high total volume of coolant needed. The coolant must obviously be fluid at the operating temperature of the reactor. Its thermal conductivity and specific heat should be high in order to achieve the maximum heat transfer from the fuel to the fluid. Further, if the coolant is a compound, it must be stable from the standpoint of dissociation by heat or neutron bombardment. Finally, the nature of the radioactivity induced by irradiation will determine the techniques necessary in handling the cycling fluid.

Many modes of cooling a reactor have been designed. In general they can be summarized as follows:

1. A fluid is passed through annular ducts surrounding, or threading, solid fuel elements.
2. A homogeneous fluid mixture of fuel and moderator is circulated through the reactor and to an external heat exchanger. The associated radioactivity problems involved in this method are very serious.
3. Convective cooling is obtained by complete immersion of the reactor fuel elements into a coolant bath.

Table 2·4   Properties of reactor coolants

A. Gaseous

	Conductivity k, Btu/hr-°F-ft	Specific heat c, Btu/lb-°F
Hydrogen......	0.089	3.4
Helium........	0.075	1.25
Air	0.014	0.25
Steam	0.012	0.45

B. Liquid

	Conductivity, Btu/hr-°F-ft	Specific heat, Btu/lb-°F	Melting point, °F	Boiling point, °F
Water	0.35	1.0	32	212
Lithium...............	22	1.0	354	2403
Sodium	50	0.33	208	1621
Sodium potassium (NaK)......	15	0.27	66	1518
Bismuth	9	0.034	520	2691
Mercury	4.8–7.3	0.033	−39	675
Lead	9.4	0.039	621	3159

Table 2·4 gives several possible coolants and a partial listing of their properties. In addition to these properties, various factors must be considered in the design of a reactor cooling system: e.g., hydrogen or deuterium gas might be suitable coolants except for the ease with which they leak from any enclosed system. If a coolant is to be passed once through a reactor and exhausted to the atmosphere, its replacement cost must be considered. Air would be a practical choice, while helium gas clearly would not, because of its scarcity. Conversely, if a closed coolant cycle is desired, helium might be preferable, because of its high thermal conductivity relative to air. Water and heavy water are useful for reactors operating at temperatures lower than their boiling points, because the steam phase is considerably less effective in removing heat.

Of the group of liquid metals, lithium and mercury are not desirable for thermal reactors because of high absorption cross sections. Liquid sodium, by virtue of its very high conductivity, would seem to be a natural choice. However, examination of the isotopes formed during irradiation shows that Na^{23} becomes Na^{24}, of long half-life (14.9 hr) with respect to the coolant cycle. This effect complicates coolant handling and must therefore be a design consideration.

2·7 Structural materials

The core of a nuclear reactor, while simple in principle, is actually a highly complex unit containing many structural components, such as supports or containers for the fuel and moderator, ductwork, piping, valves and fittings, control-rod sleeves and mechanisms, shells, baffle plates, header chambers, and closures. The general requirements of structural materials are: adequate strength, fabricability, thermal stability, radiation stability, satisfactory corrosion resistance, and desirable nuclear properties.

The number of elements suitable for structural use in thermal reactors is severely limited, since the only structural metals having cross sections lower than 0.5 barn per atom for thermal neutron capture are zirconium, beryllium, aluminum, and magnesium. Until quite recently, zirconium was an "exotic" metal which had very little commercial application. Research sponsored by the USAEC during the past fifteen years has vastly increased our information about the production, fabrication, metallurgy, and properties of zirconium. Improved metal-processing techniques, notably hafnium removal, have resulted in a product with a lower capture cross section than was originally measured. This, coupled with a high melting point, excellent fabricability, and generally good corrosion resistance, makes zirconium a promising structural material for the core of a thermal reactor. However, the pure metal is difficult and expensive to produce, and its strength at high temperature is poor. Its corrosion resistance to high-temperature water and its high temperature strength are improved considerably by additions of a few per cent of tin (zircoloy).

Beryllium is another metal which is promising for structural use, since it has a good moderating ratio as well as a high melting point (1315°C). However, it is highly toxic, it is difficult to fabricate, and its corrosion resistance is generally inferior to that of zirconium. Attempts to clad beryllium with corrosion-resistant materials have met with little success.

The remaining two low-cross-section metals, magnesium and aluminum, are certainly not high-temperature materials. Aluminum technology is well established, and in fact aluminum has been used extensively as a structural material in low-temperature research and production reactors. Although the corrosion resistance of magnesium is generally poor, it has not been accorded the attention it merits, considering its low cross section and relative abundance. Alloying aluminum or magnesium with small amounts of other materials often can bring about significant improvements in mechanical properties.

A great many more structural metals become available if a higher cross section can be tolerated. These could be used in small quantities in a thermal reactor and extensively in an intermediate or fast reactor. Pure metals, with moderate capture cross sections and high melting points, include niobium, iron, molybdenum, chromium, copper, nickel, vanadium, and titanium. Niobium has good mechanical properties and is available in high-strength weldable alloys. Chromium has not yet been developed in forms suitable for structural use. The properties of iron and copper can be greatly improved by suitable alloy additions, and so their use as pure metals is unlikely. Ductile vanadium, now available commercially, is a promising structural material from both strength and corrosion considerations. Because of its known refractory properties and unusually high thermal conductivity, molybdenum is a good possibility for high-temperature structural applications. Titanium and titanium alloys have very high strength-to-weight ratios, good corrosion resistance in general, and many desirable physical properties. Titanium technology has advanced tremendously in recent years, making titanium and its alloys prospects for structural use in intermediate reactors.

A large number of promising alloys can be produced from the pure metals of interest. Among the iron-base alloys, there are a host of stainless steels combining good corrosion resistance with good mechanical properties up to moderately high temperatures. Nickel and its alloys, such as Nichrome, Illium, Hastelloy, Monel, Inconel, Inconel X, and many others, are also interesting as structural materials.

While weak in tension and thermal-shock resistance, ceramics and graphites are the only good prospects as structural materials for very-high-temperature applications. Oxides, such as beryllia, magnesia, silica, alumina, and zirconia, are widely used as refractory materials. Of the carbides, beryllium carbide, silicon carbide (carborundum), and zirconium carbide are also prospects. In addition, there is an endless number of combinations of various oxides and carbides which can be formed to give improved properties. Furthermore, the vast new field of metal ceramics (cermets) gives promise of providing high-temperature structure materials having good strength and thermal conductivity.

2·8   Shielding materials

The basic requirements for a radiation shield around a nuclear reactor are:
1. A good moderating material, to slow down neutrons
2. A good neutron absorber

3. A dense material, for attenuating the gamma radiation

No one material answers all these demands, except when it is used in large quantities. Therefore, radiation shields generally are composed of several materials, divided roughly into the following classes:

1. Amalgams
2. Cement and concretes of special aggregates
3. Ceramics and cermets
4. Glasses and fused salts
5. Metal ores
6. Metals, alloys, and sintered powders
7. Organics, such as plastics, metal esters, metal-loaded resins, elastomers, and silicones
8. Silica and other gels precipitated from boron-loaded solutions
9. Water, hydrides, hydrates, and hydroxiders

Boron is one of the most interesting of the light elements used for neutron shielding. Although its elastic scattering cross section is low compared with hydrogen, it has an unusually high neutron capture cross section in the lower energy regions.

Hydrogenous materials are excellent moderators and also have fairly good capture cross sections for thermal neutrons. However, the capture of neutrons by hydrogen is largely an (n,γ) reaction, producing secondary gammas, which must in turn be attenuated by additional shielding. Ordinary water makes a satisfactory shield if enough thickness is used.

From a nuclear point of view, heavy metal hydrides appear to be ideal for shielding, since they could combine moderator, absorber, and gamma attenuator into a single compound, such as tungsten or tantalum borohydride. Not all metal hydrides are stable thermally, although such hydrides as titanium hydride, zirconium hydride, sodium hydride, potassium hydride, etc., might be considered stable if they were to contain somewhat less than the stoichiometric amount of hydrogen.

Concrete is a good structural material; it can be handled easily and contains enough hydrogen to moderate fast neutrons. Its density is low; consequently, an all-concrete shield will be thick. In the past few years, a great deal of development work has been done on special concretes for shields. One example of a special concrete shield is that used in the X-10 reactor at Oak Ridge. This concrete shield is made of portland cement and haydenite (a porous, calcined water-absorbent shale), coated with a bituminous paint to retard water loss. This shield has shown remarkable water retention, freedom from radiation-damage effects, and continued strength after more than five years of operation.

Lead, tungsten, and tantalum are the heavy metals of practical interest for shields. Gold, iridium, osmium, uranium, and the rare earths might be used, but the initial cost and availability are limitations. Lead and tantalum are readily fabricable, and tungsten can be made into a wide variety of shapes by powder-metallurgy techniques. All three metals oxidize readily in air at elevated temperatures and would have to be coated if service in shields is required.

3

WHAT IS A METAL?

3·1 Introduction

The materials used in the construction of nuclear reactors can be classified into two groups: metals and nonmetals. Based on the reactors built to date, the first group plays a more important role. Except for shielding materials, coolants, and possibly a graphite moderator, the remainder of most reactors is made of metals. Therefore, it is obvious that one must know something about metals in order to understand the particular requirements and problems faced in designing and fabricating nuclear reactors.

What is a metal? If you were to ask this question of three different scientists—a physicist, a chemist, and a metallurgist—all of whom were actively concerned with metals, you would receive three different answers. Each man would define a metal in terms of the particular properties or characteristics associated with his point of view.

The physicist would describe a metal in terms of the interactions of the elementary particles of matter, i.e., electrons, protons, and neutrons. The chemist would be interested in the behavior and properties of metals either in their ionized form or as part of a chemical compound. The metal ion M^{++} is formed by losing two electrons, as symbolically described in Eq. (3·1).

$$M = M^{++} + 2e^- \qquad (3·1)$$

The theoretical metallurgist likes to think of a metal as an agglomeration of closely packed atomic spheres surrounded by a sea of electrons. The more practical metallurgist would probably describe a metal in terms of such properties as plasticity, conductivity, and strength.

Thus, to secure a complete picture, we must examine all the viewpoints.

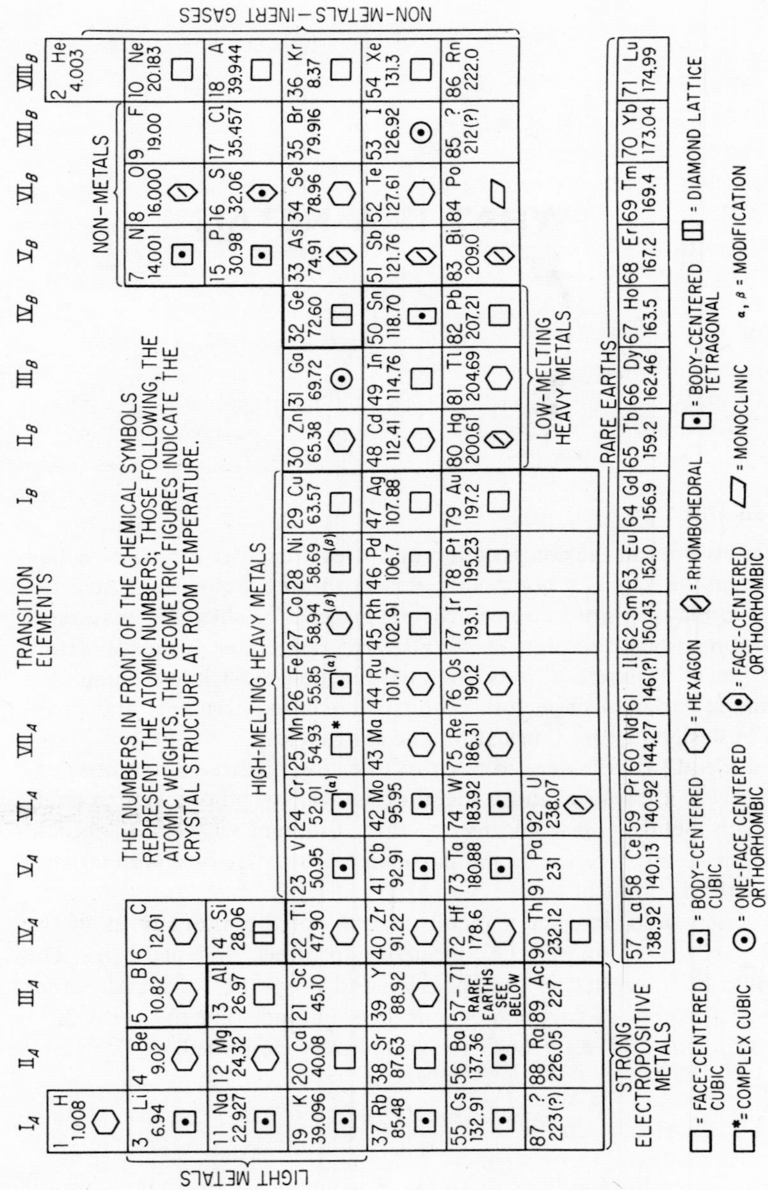

NON-METALS—INERT GASES

NON-METALS

TRANSITION ELEMENTS

HIGH-MELTING HEAVY METALS

LOW-MELTING HEAVY METALS

RARE EARTHS

STRONG ELECTROPOSITIVE METALS

LIGHT METALS

THE NUMBERS IN FRONT OF THE CHEMICAL SYMBOLS REPRESENT THE ATOMIC NUMBERS; THOSE FOLLOWING, THE ATOMIC WEIGHTS. THE GEOMETRIC FIGURES INDICATE THE CRYSTAL STRUCTURE AT ROOM TEMPERATURE.

I_A	II_A	III_A	IV_A	V_A	VI_A	VII_A	VIII			I_B	II_B	III_B	IV_B	V_B	VI_B	VII_B	VIII_B
1 H 1.008																	2 He 4.003
3 Li 6.94	4 Be 9.02	5 B 10.82	6 C 12.01											7 N 14.001	8 O 16.000	9 F 19.00	10 Ne 20.183
11 Na 22.927	12 Mg 24.32	13 Al 26.97	14 Si 28.06											15 P 30.98	16 S 32.06	17 Cl 35.457	18 A 39.944
19 K 39.096	20 Ca 40.08	21 Sc 45.10	22 Ti 47.90	23 V 50.95	24 Cr 52.01 (α)	25 Mn 54.93	26 Fe 55.85 (α)	27 Co 58.94 (β)	28 Ni 58.69 (β)	29 Cu 63.57	30 Zn 65.38	31 Ga 69.72	32 Ge 72.60	33 As 74.91	34 Se 78.96	35 Br 79.916	36 Kr 8.37
37 Rb 85.48	38 Sr 87.63	39 Y 88.92	40 Zr 91.22	41 Cb 92.91	42 Mo 95.95	43 Ma 146(?)	44 Ru 101.7	45 Rh 102.91	46 Pd 106.7	47 Ag 107.88	48 Cd 112.41	49 In 114.76	50 Sn 118.70	51 Sb 121.76	52 Te 127.61	53 I 126.92	54 Xe 131.3
55 Cs 132.91	56 Ba 137.36	57–71 RARE EARTHS SEE BELOW	72 Hf 178.6	73 Ta 180.88	74 W 183.92	75 Re 186.31	76 Os 190.2	77 Ir 193.1	78 Pt 195.23	79 Au 197.2	80 Hg 200.61	81 Tl 204.69	82 Pb 207.21	83 Bi 209.0	84 Po	85 ? 212(?)	86 Rn 222.0
87 ? 223(?)	88 Ra 226.05	89 Ac 227	90 Th 232.12	91 Pa 231	92 U 238.07												

RARE EARTHS

57 La 138.92	58 Ce 140.13	59 Pr 140.92	60 Nd 144.27	61 Il 146(?)	62 Sm 150.43	63 Eu 152.0	64 Gd 156.9	65 Tb 159.2	66 Dy 162.46	67 Ho 163.5	68 Er 167.2	69 Tm 169.4	70 Yb 173.04	71 Lu 174.99

Legend:

□ = FACE-CENTERED CUBIC
□* = COMPLEX CUBIC
□• = BODY-CENTERED CUBIC
□⊙ = ONE-FACE CENTERED ORTHORHOMBIC
⬡ = HEXAGON
◇ = RHOMBOHEDRAL
⊙ = FACE-CENTERED ORTHORHOMBIC
◇ = MONOCLINIC
▭ = BODY-CENTERED TETRAGONAL
⬒ = DIAMOND LATTICE
α, β = MODIFICATION

Fig. 3·1   The periodic system of the elements.

18

PART I. MODERN PHYSICS AND CHEMISTRY

3·2 The elements

All matter consists of atoms. Briefly, it is known that every atom consists of a small, positively charged nucleus about which are swarming negatively charged particles called electrons. The nature of an atom is determined uniquely by the charge on the nucleus, which is always an integer multiple of a fundamental unit e equal to 1.6×10^{-9} coulomb. (A coulomb is the amount of electricity that flows by a point in a circuit in 1 sec when the current is 1 amp.) This integer, which is called the atomic number, generally increases with the atomic weight of the element, being equal to unity for the lightest element, hydrogen, and to 92 for the heaviest natural element, uranium. The electrons moving about the nucleus are identical and have charge $-e$. In the normal state of the atom, they are equal in number to the atomic number of the atom. Thus, the atom is electrically neutral in its normal state.

In the early days of the development of chemistry, it was discovered that the properties of elements vary periodically when plotted as functions of the atomic weight. This periodic behavior has given rise to the periodic chart of the elements given in Fig. 3·1. Here the elements appear in a sequence of increasing atomic number running from left to right. The chart shown in Fig. 3·1 has been divided to illustrate several of the periodic features of the table. Groupings of the light metals, high- and low-melting heavy metals, nonmetals, inert gases, etc., are given in brackets.

3·3 The atom

With the development of the modern picture of atomic structure the periodic chart has received complete interpretation. The essential difference between atoms lies in the number of electrons circulating about the nucleus of each. The exact nature of the motion of the electrons about the nucleus can be de-

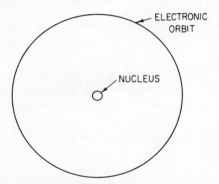

Fig. 3·2 Schematic illustration of a hydrogen atom.

scribed only in mathematical language more formidable than is necessary for our purpose. The field of quantum mechanics concerns itself with this problem of position and energy of the electrons. In an approximate, though adequate, way, it can be said that the electrons revolve about the nucleus in orbits resembling somewhat the orbits of the planets about the sun (see Fig. 3·2). In the simplest case, hydrogen, there is a single electron moving in a single orbit, whereas, in the most complicated naturally occurring atom, uranium, there are 92 electrons moving in about half that number of different orbits. There are, however, important differences, apart from size, between electronic orbits and planetary orbits.

The energy of any planet, though fixed, could be altered, at least in principle, by an arbitrary amount. The energies of the orbital electrons in atoms, however, are restricted to definite, or discrete, values. For example, it is possible for a planet to move in any elliptical orbit that might be drawn with the sun at one focus, whereas only a comparatively few such orbits are allowed in the atomic case. The discreteness of energy states is characteristic of the quantum-physics approach to all problems dealing with atomic systems.

3·4 Review of quantum numbers

The simplest way to discuss electronic orbits is in terms of the energy-level diagram, in which the energies of the allowed states of motion, or

Table 3·1 Energy levels for the elements

Number of electrons in each level	Conventional designations for the levels	
10	$6d$	(6,2)
6	$7p$	(7,1)
2	$7s$	(7,0)
14	$7f$	(4,3)
10	$5d$	(5,2)
6	$6p$	(6,1)
2	$6s$	(6,0)
10	$4d$	(4,2)
6	$5p$	(5,1)
2	$5s$	(5,0)
10	$3d$	(3,2)
6	$4p$	(4,1)
2	$4s$	(4,0)
6	$3p$	(3,1)
2	$3s$	(3,0)
6	$2p$	(2,1)
2	$2s$	(2,0)
2	$1s$	(1,0)

Energy $\longrightarrow$

orbits, are represented as horizontal lines on a vertical energy scale (see Table 3·1). It is customary to associate with each level two positive integers n and l, which are related, respectively, to the energy of the electron and to its angular momentum relative to the nucleus. A more detailed discussion of these two quantum numbers has been presented previously.

The allowed values of *n* range from 1 to ∞ and those of *l* from 0 to *n* − 1 for a given *n*. Thus, if we use the notation (*n,l*) to designate the pairs of numbers characterizing a given level, for *n* equal to 1, there is a (1,0) level, but no (1,1) level. Similarly, for *n* equal to 2, there are (2,0) and (2,1) levels, but no (2,2) or (2,3) levels. In general, the orbits for which *n* and *l* are smallest lie nearest the nucleus and have the lowest energy, as is shown in Table 3·1.

During the early research upon atomic structure, different investigators introduced different ways of designating the orbits. Though all such schemes of notation are essentially equivalent, so that only the one given above is necessary, another is used so frequently by metallurgists and physicists that we shall present it here. In this scheme, the integers *l* are replaced by letters in accordance with Table 3·2. Thus, the level designated by (2,0) in the

Table 3·2

l	0	1	2	3	4	5	6
Letter......	*s*	*p*	*d*	*f*	*g*	*h*	*i*

first scheme is also designated by 2*s*. Both schemes of notation are used in Table 3·1.

3·5 Details of the periodic table

Let us consider the manner in which the levels are occupied by electrons when we consider the elements in a sequence of increasing atomic number, as they appear in the periodic chart. The first element, hydrogen, has one electron, which occupies the lowest, or 1*s*, level. Helium has two electrons, both of which occupy the 1*s* level. If more electrons could enter this level, lithium would have three 1*s* electrons. Actually, it is found that only a limited number of electrons can occupy any given atomic orbit. This restriction is a special case of a general rule, called the Pauli exclusion principle in honor of its discoverer. According to this rule, the maximum number of electrons that can occupy an orbit of given *l* is $2(2l + 1)$. The numbers of electrons $2(2l + 1)$ are known as the degeneracies of the levels. Thus, we see that only two electrons are allowed in the 1*s* state and that the third electron of lithium must enter a level of higher energy. Actually, it enters the next lowest, or 2*s*, level. The inert characteristics of helium are associated with the fact that its two electrons occupy completely the 1*s* level, which is widely separated from the 2*s* and 2*p* group. A completed group of levels that is widely separated from other groups is usually a closed shell.

Beryllium possesses two electrons in the 2*s* level. This element is not a rare gas, however, because of the proximity of the 2*p* level. Only when we reach neon, in which both the 2*s* and 2*p* levels are filled, do we again obtain

Table 3·3 Details of electronic structure of the elements

Element and atomic number	Principal and secondary quantum numbers									
$n =$	1	2	2	3	3	3	4	4	4	4
$l =$	0	0	1	0	1	2	0	1	2	3
1 H	1									
2 He	2									
3 Li*	2	1								
4 Be*	2	2								
5 B	2	2	1							
6 C	2	2	2							
7 N	2	2	3							
8 O	2	2	4							
9 F	2	2	5							
10 Ne	2	2	6							
11 Na*	2	2	6	1						
12 Mg*	2	2	6	2						
13 Al*	2	2	6	2	1					
14 Si?	2	2	6	2	2					
15 P	2	2	6	2	3					
16 S	2	2	6	2	4					
17 Cl	2	2	6	2	5					
18 A	2	2	6	2	6					
19 K*	2	2	6	2	6		1			
20 Ca*	2	2	6	2	6		2			
21 Sc*	2	2	6	2	6	1	2			
22 Ti*	2	2	6	2	6	2	2			
23 V*	2	2	6	2	6	3	2			
24 Cr*	2	2	6	2	6	5	1			
25 Mn*	2	2	6	2	6	5	2			
26 Fe*	2	2	6	2	6	6	2			
27 Co*	2	2	6	2	6	7	2			
28 Ni*	2	2	6	2	6	8	2			
29 Cu*	2	2	6	2	6	10	1			
30 Zn*	2	2	6	2	6	10	2			
31 Ga*	2	2	6	2	6	10	2	1		
32 Ge*	2	2	6	2	6	10	2	2		
33 As?	2	2	6	2	6	10	2	3		
34 Se	2	2	6	2	6	10	2	4		
35 Br	2	2	6	2	6	10	2	5		
36 Kr	2	2	6	2	6	10	2	6		

Element and atomic number	Principal and secondary quantum numbers										
$n =$	1	2	3	4	4	4	4	5	5	5	6
$l =$	–	–	–	0	1	2	3	0	1	2	0
37 Rb*	2	8	18	2	6			1			
38 Sr*	2	8	18	2	6			2			
39 Yt*	2	8	18	2	6	1		2			
40 Zr*	2	8	18	2	6	2		2			
41 Nb*	2	8	18	2	6	4		1			
42 Mo*	2	8	18	2	6	5		1			
43 Ma*	2	8	18	2	6	6		1			
44 Ru*	2	8	18	2	6	7		1			
45 Rh*	2	8	18	2	6	8		1			
46 Pd*	2	8	18	2	6	10		–			
47 Ag*	2	8	18	2	6	10		1			
48 Cd*	2	8	18	2	6	10		2			
49 In*	2	8	18	2	6	10		2	1		
50 Sn*	2	8	18	2	6	10		2	2		
51 Sb?	2	8	18	2	6	10		2	3		
52 Te	2	8	18	2	6	10		2	4		
53 I	2	8	18	2	6	10		2	5		
54 Xe	2	8	18	2	6	10		2	6		
55 Cs*	2	8	18	2	6	10		2	6		1
56 Ba*	2	8	18	2	6	10		2	6		2
57 La*	2	8	18	2	6	10		2	6	1	2
58 Ce*	2	8	18	2	6	10	1	2	6	1	2
59 Pr*	2	8	18	2	6	10	2	2	6	1	2
60 Nd*	2	8	18	2	6	10	3	2	6	1	2
61 Il*	2	8	18	2	6	10	4	2	6	1	2
62 Sm*	2	8	18	2	6	10	5	2	6	1	2
63 Eu*	2	8	18	2	6	10	6	2	6	1	2
64 Gd*	2	8	18	2	6	10	7	2	6	1	2
65 Tb*	2	8	18	2	6	10	8	2	6	1	2
66 Ds*	2	8	18	2	6	10	9	2	6	1	2
67 Ho*	2	8	18	2	6	10	10	2	6	1	2
68 Er*	2	8	18	2	6	10	11	2	6	1	2
69 Tm*	2	8	18	2	6	10	12	2	6	1	2
70 Yb*	2	8	18	2	6	10	13	2	6	1	2
71 Lu*	2	8	18	2	6	10	14	2	6	1	2
72 Hf*	2	8	18	2	6	10	14	2	6	2	2

* Denotes metals.

Table 3·3 (Cont.)

n =	1	2	3	4	5			6			7
l =	-	-	-	-	0	1	2	0	1	2	0
73 Ta*	2	8	18	32	2	6	3	2			
74 W*	2	8	18	32	2	6	4	2			
75 Re*	2	8	18	32	2	6	5	2			
76 Os*	2	8	18	32	2	6	6	2			
77 Ir*	2	8	18	32	2	6	7	2			
78 Pt*	2	8	18	32	2	6	8	2			
79 Au*	2	8	18	32	2	6	10	1			
80 Hg*	2	8	18	32	2	6	10	2			
81 Tl*	2	8	18	32	2	6	10	2	1		
82 Pb*	2	8	18	32	2	6	10	2	2		
83 Bi?	2	8	18	32	2	6	10	2	3		
84 Po	2	8	18	32	2	6	10	2	4		
85 —	2	8	18	32	2	6	10	2	5		
86 Em	2	8	18	32	2	6	10	2	6		
87 —	2	8	18	32	2	6	10	2	6		1
88 Ra*	2	8	18	32	2	6	10	2	6		2
89 Ac*	2	8	18	32	2	6	10	2	6	1	2
90 Th*	2	8	18	32	2	6	10	2	6	2	2
91 Pa*	2	8	18	32	2	6	10	2	6	3	2
92 U*	2	8	18	32	2	6	10	2	6	4	2

* Denotes metals.

a rare gas. The next element, sodium, like lithium, has one electron outside of a closed shell. This fact is intimately related to the similar chemical properties of the two elements. There is a one-to-one correspondence between the sequence of elements ranging from magnesium to argon and the sequence ranging from beryllium to neon in which the $3s$ and $3p$ levels are filled. The element following argon, namely, potassium, is again an alkali metal. The details of the electron structure of free atoms are given in Table 3·3.

In the sequence between potassium and the next rare gas, krypton, we see that a new type of element enters the picture. At this point, the $3d$ shell, which lies near the $4s$ and $4p$ shells as shown in Table 3·1, begins to be occupied. The elements having partially filled d shells are called the transition metals. The property of ferromagnetism is demonstrated by several of the transition group which includes iron, cobalt, and nickel. As the $3d$ level fills, its energy level drops relative to the $4p$ and $4s$ levels for reasons too complex to discuss here. Thus, the element nickel, with 10 electrons outside the argon group of shells, has the $3d$ level below the $4p$ and $4s$ levels. The next element, copper, has 1 electron in the 4s level and, as a result, has properties that resemble in part those of alkali metals. With the filling of the 4s and 4p shells, we reach the rare gas krypton.

The sequence from potassium to krypton repeats itself in the period extending from rubidium to xenon in which the 5s, 5p, and 4d levels become occupied. By combining the information presented in Table 3·1 and Table 3·3 we can follow the further details of the periodic table.

Beyond xenon, the sequence from rubidium to xenon essentially repeats itself, although the 4f shell of 14 electrons is filled just after lanthanum, giving rise to the 14 rare-earth elements. The chemical and physical properties of these elements are very similar in that they differ only with regard to the number of comparatively inert 4f electrons they possess.

The atomic nuclei beyond lead are relatively unstable. The instability is made evident by the presence of radioactive elements in this sequence. The last naturally occurring element, uranium with atomic number 92, apparently results from complete instability of elements of higher atomic number. In recent years, by use of particle accelerators and nuclear reactors, elements up to atomic number 100 have been produced for short periods of time.

3·6 Characteristics of metals

Now that we have examined the components and structure of all the elements, what can we say about the metals? If we reexamine Fig. 3·1 and notice the elements which have been designated metals, several interesting

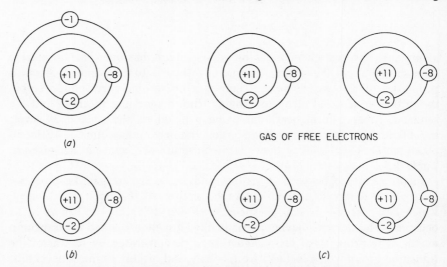

Fig. 3·3 Schematic illustration of electronic levels for the sodium atom, ion, and metal. (a) Neutral atom of sodium; (b) positively charged sodium ion; (c) a sodium crystal, showing the closed shells about each sodium ion and the free-electron gas.

facts are illuminated. All the metals, with the possible exception of arsenic and antimony (which are borderline metals, at best), contain either one or two electrons in the outermost electronic shell. This characteristic of metal atoms plays an important role in the combining of single atoms to form gaseous, liquid, or solid matter.

When atoms are brought together in large groups to form gases, liquids, or solids, each individual atom prefers to arrange itself, with respect to its neighbors, so that its outer electronic shell is completely filled.

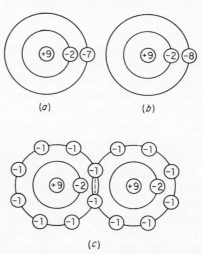

As an example of this behavior of atoms, let us examine the metal atom sodium and the nonmetal atom fluorine. Sodium, with an atomic number of 11, has the electrons arranged in shells of 2, 8, and 1. The one electron in the outer shell is known as the valence electron. Figure 3·3a represents a schematic arrangement of electrons for the sodium atom. When the sodium atom is brought into contact with other atoms, it prefers to ionize, i.e., to lose an electron and become a positively charged particle, as shown in Fig. 3.3b. The positive

Fig. 3·4 Schematic illustration of electronic levels for the fluorine atom, ion, and molecule. (a) Neutral atom; (b) negatively charged ion; (c) fluorine molecule.

charge arises from a charge of $+11$ in the nucleus and only 10 electrons surrounding the nucleus. Ionization of sodium can be represented by Eq. (3·2).

$$Na \rightarrow Na^+ + 1e^- \tag{3·2}$$

Fluorine, on the other hand, with an atomic number of 9, has its electrons arranged in shells of 2 and 7. The fluorine atom needs only one more electron to form the closed shell and become stable. Figure 3·4a represents a schematic arrangement of the electrons in the neutral fluorine atom. When fluorine is added to any group of atoms, it ionizes by gaining an electron.

$$F + 1e^- \rightarrow F^- \tag{3·3}$$

Thus, we again see the preference for each atom to cause its outermost electronic shell to be completely filled. Metal atoms accomplish this condition by losing electrons, while nonmetal atoms gain electrons to achieve the same end.

One further example of the difference between metal and nonmetal atoms is furnished by the mechanism of forming clusters of atoms into gases or solids. Figure 3·3c schematically illustrates how the solid metal, sodium, is formed. The cohesive forces in the sodium crystal are due to a "gas" of electrons, free to move among the positive metallic ions in definite energy levels, holding together the positive metallic ions by virtue of their mutual attraction for the negative electron gas. An enlargement of these concepts would involve the introduction of the "free-electron" and "bond" theory of metals, topics much beyond the scope of this text.

Figure 3·4c represents the binding forces in a fluorine (F_2) molecule. The binding forces are of a covalent nature in which neighboring atoms share valence electrons in order to build up stable electron shells. The example in the figure shows two fluorine atoms with a valence of −1 sharing one of the seven electrons in their outer shells with each other, thus building up a stable external shell and forming a fluorine molecule.

Thus, we see how the number of electrons in the outer shell can effect the manner in which atoms group together to form matter. It is here that the "solid-state physicist" takes up the challenge. What are the binding forces that hold atoms together? How do these forces control the chemical and physical properties of the aggregate? How can we calculate quantitatively these binding energies and thus predict the chemical and physical properties? These are some of the important questions facing investigators of the solid state. Progress is very rapid at present, and many properties of matter can be described and predicted from basic fundamentals. A few of these properties are magnetic moment, ionization potential, crystallographic forms, and elastic moduli.

3·7 Summary

The point of view of the physicist concerning the question What is a metal? can be summarized as follows: The physicist likes to think of matter in terms of its fundamental building blocks, such as neutrons, protons, and electrons. Of all matter, he finds that metals always have a small number of electrons in their outer orbit, compared with the number required to complete a stable configuration, or shell. Furthermore, he finds that metal atoms group together in a particular way so that the valence, or outer, electrons form a sort of "electron gas" about the positively charged metal ions. This electron gas is responsible for the characteristic metallic properties of metals.

3·8 Modern chemistry

Modern chemistry, naturally, takes advantage of all the advances in physics and mathematics to solve its particular problems. Therefore, it is not unusual to find chemists working with the same techniques as those used by solid-state physicists. The chemist would, however, probably slant the choice of problems to the allied fields of organic and inorganic substances.

In a crude, but helpful, manner, most chemistry can be broken down into two components: organic and inorganic. The name organic is historic in origin and refers to the compounds of carbon. When it was discovered that both plant and animal tissue is made up chiefly of carbon compounds, it was thought that such compounds could be synthesized only through the aid of living matter; thus, they were called organic compounds. Although it has since been learned that no such limitation exists, chemists continue to designate the chemistry of the carbon compounds as organic chemistry.

Inorganic chemistry, therefore, covers the chemistry of all the other elements and compounds. Since some 75 of the 92 naturally occurring elements are metals, the inorganic chemist and the metallurgist would appear to have similar interests. In some cases, this is true, but, for the most part, chemists are interested in chemical properties such as ionization, reaction rates, vapor pressure, etc., while metallurgists are interested in physical properties such as tensile strength, elastic modulus, thermal and electrical conductivity, etc.

Now we shall return to the point in question, i.e., how a chemist would describe a metal. Probably his first answer would be based on the electrolytic behavior of metals, as contrasted with the behavior of nonmetals. This criterion classifies as a metal any element which upon electrolysis of its compounds appears at the cathode of an electrolytic cell. The nonmetals, such as chlorine and oxygen, would go to the anode. Obviously, this classification is based on the fact that metals become positively charged when ionized and move toward the cathode, which is negatively charged. Thus, we see that the chemist's description is simply a practical way of phrasing the physicist's point of view: briefly, metals have either one or two outer electrons which they would like to lose, whereupon they would become positively charged ions with a completed outer orbit.

For the most part, chemists are interested, not in metals as such, but in metallic ions, compounds, or radicals. However, a discussion of these subjects would require a complete course in itself. We shall now examine the point of view of the metallurgist, for as his title implies he is primarily interested in the metallic elements as part of a liquid or solid phase.

PART 2. METALLURGY

3·9 Attitude of a metallurgist

The modern metallurgist could be called an "empirical solid-state physicist." That is, the metallurgist, though concerned with the solid state of metals, has a different point of view from that of the physicist. Whereas the solid-state physicist would concentrate on how the interactions between electrons explain the properties of metals, the metallurgist seeks a quantitative evaluation of these properties and develops methods to control and improve them.

3·10 Properties of metals

The combination of properties possessed by metals, by virtue of which they derive their preeminence in mechanical and structural uses, is essentially the combination of strength with plasticity. Many brittle substances, glass, for instance, have high tensile strengths, as high as 400,000 psi, but they lack plasticity under sudden shock (as anyone knows who has dropped his wife's best china plate). Glass, tar, sealing wax, and even rocks (in the earth under tremendous pressures and over periods of millions of years) can be made to flow slowly. But metals can be deformed rapidly, without fracture, by sudden shock, as in stamping (e.g., coin manufacturing) and forging (e.g., automobile crankshafts). A bridge made of glass might have ample strength, but if not perfectly aligned, it would break. A metal bridge, on the other hand, could conform to a considerable change of alignment by both elastic and plastic deformation without rupture, in fact without loss of strength.

Corrosion resistance is another of the more useful properties of metals. Resistance to attack by chemical agents may be sufficiently great for the metal to resist hot acids or corrosive gases at elevated temperatures, or corrosion resistance may be so low that the metal barely resists exposure to the atmosphere. A common example of the latter condition is the rusting of unprotected iron. A common example in the USAEC laboratories is the tarnishing of uranium and the sudden fires which start spontaneously in zirconium and uranium chips. Both zirconium and uranium are very reactive materials; i.e., these materials have a great tendency to combine with oxygen at low temperatures. In the case of zirconium, a thin layer of zirconium oxide (ZrO_2) surrounds the metal and protects the remainder from further oxidation as long as the oxide skin remains intact.

Some metals or alloys are valued for aesthetic reasons; gold, silver, platinum, bronze, and stainless steel are good examples. Most metals or alloys used for their good appearance have good corrosion resistance. Improved appearance and resistance to corrosion can often be imparted to baser (more reactive) metals by treating their surfaces chemically or by coating them with paints, enamels, or thin films of nobler (less reactive) metals.

A unique property of metals is high electrical conductivity. Copper, silver, and aluminum are among the best conductors, preference depending on which metal best and most economically meets other service requirements. Another unique property of metals is thermal conductivity; in general, metals that conduct electricity easily also conduct heat well.

At 'this point, we shall digress a little to point out a connection between the properties of metals as described above and the theoretical model of a metal developed in the first section of this chapter. If you will remember the

Table 3·4 Special properties of metals, typical examples, and uses

Property	Metals	Use
Ferromagnetism	Fe, Ni, Co, Ge	Transformers, magnets
Thermionic (electron) emission	Cu, Cr	Vacuum tubes
Optical reflectivity...........	Ag, Pt	Mirrors
High melting point	W, Mo	Filaments in incandescent lamps
Converting thermal energy into electricity	Cu, Fe, Pt	Thermocouples (tempera- ture measurement)
Magnetostriction and piezoelectricity	Ni, Co, Fe, Si	Ultrasonics

simple model of a metal presented, i.e., positive ions surrounded by a gas of free electrons, we shall try to explain why metals are good conductors of heat and electricity. When an electric field or a thermal gradient is applied to matter, it affects both the atom's nucleus and its outer electrons. The transfer of heat is carried out by two processes, first by vibration of the nuclei and second by motion of electrons. The second process is much more capable of passing large quantities of energy, and therefore we should expect that matter with relatively "loose," or "free," electrons would conduct heat energy at a greater rate. Experiments bear this theory out, since metals are the best conductors. For example, the thermal conductivity of silver is 1,000 times higher than that of solid iodine crystals.

The ability of matter to carry electricity is related to the motion of electrons, and once again metals are superior conductors because of the free electrons. Silver is 1 million million (10^{12}) times a better conductor of electricity than crystalline iodine.

Some special service conditions require properties for which particular metals are outstanding. Some of these properties are given in Table 3·4, together with common uses. This is only a partial list of properties; others are only beginning to be exploited, and some are still unused and unknown.

3·11 Characteristics of metals

Metals can be characterized by their behavior under certain conditions, as well as by their physical properties. Certain of these behavior traits are demonstrated by all metals and will further help answer the question: What is a metal? The characteristics we shall study should help answer these questions:

1. Do metal atoms group together in particular ways to form an overall structure?
2. How do metals deform?
3. How does heat-treatment change the properties of metals?

3·12   Structure of metals

The behavior of solids, liquids, and gases is mainly due to the way in which their atoms are bound together. In metals and in many other solids, the atoms are arranged in regular arrays called crystals. A crystal consists of atoms arranged in a pattern that repeats periodically in three dimensions. The regular repetition of the unit structure (the unit cell) is similar to a wallpaper pattern. The design on the wallpaper consists of a fundamental unit of the design placed in parallel fashion at each point of a two-dimensional lattice. The structure of a crystal consists of a unit cell, an atom or group of atoms, placed in parallel fashion at each point of a three-dimensional lattice.

A metal sample, such as a copper wire, contains many crystals, each of which is an individual, regular periodic array of atoms. The geometry of the array is called the crystal structure of the metal, and different metals solidify in different atomic patterns. Fortunately, only three different patterns, and these of the simplest type, are encountered in the most common metals, and they can be understood easily on the basis of elementary geometric relationships. However, before we discuss the three common lattice structures, how do we know that the atoms are arranged in these particular lattices? Visual examination of a copper wire cannot, of course, show that the atoms in each of its crystals are arranged in what is known as a face-centered cubic crystal structure, but if a beam of X rays is aimed at the wire, diffraction of the beam is observed and the crystal structure can be deduced. X-ray techniques are too involved to be discussed at this point.

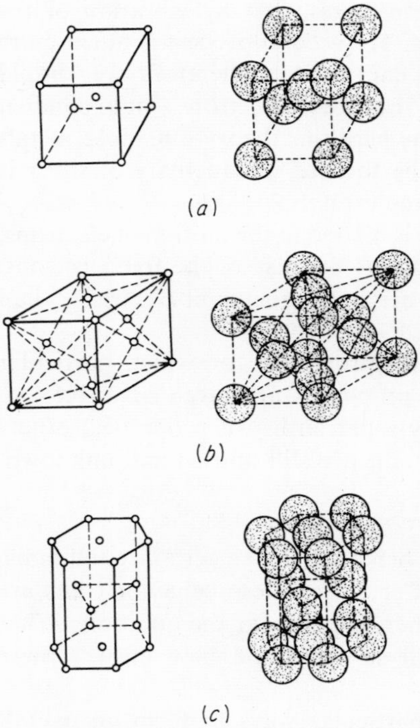

Fig. 3·5 Three common metal crystal structures. (a) Body-centered cubic; (b) face-centered cubic; (c) hexagonal close-packed.

3·13   Common crystal structures

The three most common structures found among metals have body-centered cubic (bcc), face-centered cubic (fcc), or hexagonal close-packed (hcp) lattices. These structures are shown diagrammatically in Fig. 3·5. At each corner of these units is a lattice site, which is the equilibrium position for the center of an atom.

Additional lattice sites lie within the structure outlined by the corner atoms. The spacings between corner atoms are called the lattice parameters. For cubic structures only one lattice parameter is necessary, but for hexagonal structures two are necessary, one for the spacing between the atoms in the basal planes (a) and one for the spacing between basal planes (c). The typical values for lattice parameters lie between 2 and 3 Å (1 Å equals 10^{-8} cm, or about four-billionths of an inch).

No metal actually crystallizes in the simple cubic pattern, i.e., as indicated in Fig. 3·5a, without the atom at the center of the cube. Metals such as α–Fe (ferrite), Cr, V, Mo, and W possess bcc structures. These bcc metals have two properties in common, high strength and low ductility (property that permits permanent deformation). Face-centered cubic metals such as γ–Fe (austenite), Al, Cu, Pb, Ag, Au, Ni, Pt, and Th are, in general, of lower strength and higher ductility than the bcc metals. Hexagonal close-packed structures are found in Be, Mg, Zn, Cd, Co, Ti, and Zr.

3·14 Crystalline polymorphy

Variation in temperature or pressure of a crystal, without melting or vaporization, may cause it to change its crystal structure, i.e., the form of the internal arrangement of its atoms. The ability of a material to exist in more than one crystalline form is called polymorphy. Examples of polymorphy occur in many of the materials of interest to reactor designers. A few of these examples are listed below:

1. Heating iron to 907°C causes a change from bcc (alpha, ferrite) iron to the fcc (gamma, austenite) form.
2. Zirconium is hcp (alpha) up to 863°C, when it transforms to beta zirconium, bcc.
3. Uranium exhibits three crystalline forms:
   a. Alpha (α) uranium is orthorhombic and stable up to about 666°C.
   b. Beta (β) uranium is tetragonal and stable between 666 and 760°C.
   c. Face-centered cubic gamma (γ) uranium exists from 760°C up to the melting point.

The properties of one polymorphic form of the same metal will differ from those of another form. For example, gamma iron can dissolve up to 1.7 per cent carbon, whereas alpha iron can dissolve only 0.03 per cent.

3·15 Close-packed structures

In the present state of our knowledge, the reason for a given metal's solidifying in any one of the three systems is unknown. However, it is known that each metal will try to arrange its atoms so that a condition of lowest potential energy and lowest electrostatic attraction between the atoms exists. Those metals which solidify in the fcc system or the hcp system actually attain that condition by the closest packing of the atoms. This

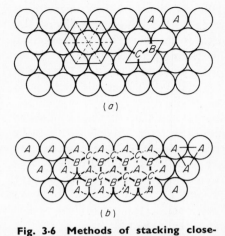

(a)

(b)

Fig. 3·6 Methods of stacking close-packed spheres. (a) Close-packed hard spheres. (b) Dotted lines indicate second layer in B position. The C position is also shown.

condition may be compared with the packing of hard spheres. If a layer of spheres is laid down as closely as possible, the appearance will be shown by the full circles in Fig. 3·6a, revealing a centered hexagon of seven spheres or an equilateral triangle connecting the centers of three spheres. In placing the second layer on top of the first, each sphere in Fig. 3·6b lies in a hollow between three spheres which form a triangle in the first layer. It should be noticed that only every alternate triangle (those marked B in Fig. 3·6a) in the first layer can be covered by the second layer. The third layer can then be laid down in two ways.

1. Each sphere may be laid vertically over a sphere in the first layer (A packing). The packing could then be represented by:

$$\begin{array}{ccccccc} A & A & A & A & A & A & A \\ B & B & B & B & B & B & B \\ A & A & A & A & A & A & A \\ B & B & B & B & B & B & B \end{array}$$

This method of packing results in the hcp arrangement.

2. Each sphere in the third layer may be laid over the centers of the triangles marked C in the first layer, which were skipped in laying the second layer. This structure could then be represented by:

$$\begin{array}{ccccccc} A & A & A & A & A & A & A \\ B & B & B & B & B & B & B \\ C & C & C & C & C & C & C \\ A & A & A & A & A & A & A \\ B & B & B & B & B & B & B \\ C & C & C & C & C & C & C \end{array}$$

This arrangement results in the face-centered cubic structure.

The bcc structure cannot be analyzed in this manner, since it cannot be represented by close-packed spheres.

3·16 Uranium crystal structure

The structure of uranium will be discussed briefly, for two reasons: first, because of its importance in the nuclear power field; second, to illustrate one of the many complex crystal structures. Uranium is orthorhombic at room temperature, with lattice parameters $a_0 = 2.854$ Å, $b_0 = 5.867$ Å, and $c_0 = 4.957$ Å, as shown in Fig. 3·7a. The uranium atom positions in the orthorhombic structure fall, not at the corners of the unit cell as in the examples discussed previously, but at the irrational positions shown in Fig. 3·7b. This structure can be considered as a distorted hcp structure in which 4 of the 12 nearest neighbors of the hexagonal array are moved in to appreciably closer distances.

The condition of orderly internal arrangement of the crystal is the essential condition of the crystalline state. Although we are accustomed to think of crystals as bodies with regular symmetrical boundary planes, we now realize that external symmetry is merely evidence of something more fundamental, i.e., regular arrangement of atoms within the crystal. Faces on a crystal and planes within a crystal are always sheets of atoms, and all edges on a crystal are lines of atoms. The physical and chemical characteristics of metals, such as coefficient of expansion, conductivity, solubility, and strength, are determined not only by the composition of the material but also by the arrangement of the atoms, i.e., by the space lattice.

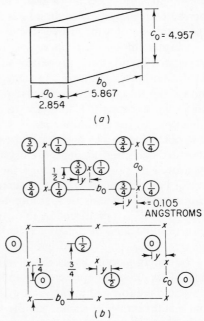

Fig. 3·7 Uranium crystal structure. (a) Orthorhombic unit cell of uranium. (b) Circled fractions are the coordinates of the atom in the perpendicular direction not shown in each diagram.

3·17 Grain structure

If we were to take a small section of a common metal and examine it under a microscope, a structure similar to that shown in Fig. 3·8a would be revealed. Each of the light areas are grains or crystals, and these are surrounded by dark lines, which are the grain or crystal boundaries. Within each grain, the atoms are arranged in a lattice structure, as described in the previous section. The grain boundaries separate differently oriented regions in which the crystal structures are identical. Figure 3·8b schematically represents four grains of different orientation and the grain boundaries which arise at the interfaces between the grains. The grain boundary is a region of misfit between grains, about one to three atom diameters wide.

One very important feature of a polycrystalline metal as pictured in Fig. 3·8a is the average size of the grains. The grain size determines many important properties of the metal; for example, smaller grain size increases the tensile strength, and large grain size is preferred for improved high-temperature creep properties. In a later section, we shall see how we can control the grain size of a metal by thermal and mechanical treatment.

Another important property of the grains is *preferred orientation*. In the vapor and liquid states, properties such as conductivity of heat and electricity or diffraction of light are the same in whatever direction through the substance the measurements of these properties are made. Substances in these states have properties that are said to be *isotropic*, i.e., independent of direction. In the crystalline state, with its geometric arrangement of atoms, it is not surprising to find that these properties vary if the direction in which they are measured through the crystal varies. For instance, if a property is measured in a direction through the crystal parallel to the layers of atoms, it will be different from the measurements made at an angle to these layers. For example, crystals of graphite conduct an electric current one hundred times better in one direction than in a direction perpendicular to the first direction. Because of this directional behavior, crystals are said to be *anisotropic*.

Returning to our discussion of preferred orientation, let us examine Fig. 3·9a. This drawing represents a random arrangement of the grains such that no one direction within the grains is aligned with the external boundaries of the metal sample. If such a sample were rolled sufficiently in one direction, it might develop a grain-oriented structure as shown in Fig. 3·9b. This structure shows a preferred orientation in the rolling direction. In many instances, preferred orientation is very desirable, but, in other cases, it can be most harmful. As we shall see in a later chapter, preferred orientation in

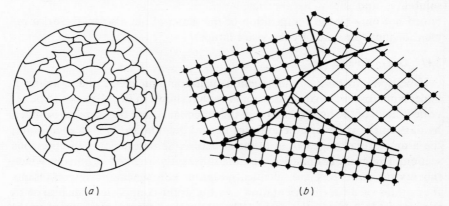

(a)   (b)

Fig. 3·8 Schematic representation of grains and grain boundaries. (a) Microscopic scale; (b) atomic scale.

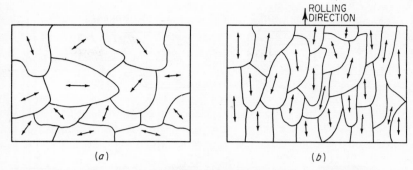

(a)

(b)

Fig. 3·9 Schematic representation of random and preferred orientation of grains. (a) Random orientation of grains; (b) preferred orientation.

uranium fuel elements can result in catastrophic changes in dimensions during use in a reactor.

3·18 Crystal planes and Miller indices

Before an examination of the mechanisms and effects of deformation of metals, let us try to understand the significance of planes in a crystal lattice. We shall have frequent cause to consider particular crystallographic planes and directions in the simpler types of lattice. The regular arrangement of atoms throughout the crystal makes it possible to pass planes in different directions through the crystal, and on these planes the atomic nuclei will be more or less densely distributed. The locations of the planes and directions in a crystal are most conveniently described by referring them to the three coordinate axes parallel to the edges of a unit cell of the lattice. If these happen to be orthogonal, as in cubic crystals, the axes are a cartesian set, as shown in Fig. 3·10a. Next, units of length along these axes are chosen equal to the corresponding lengths of the edges of the unit cell. A plane is defined by the length of its intercepts on the three axes, measured from the origin (center of coordinates). For simplicity's sake, the reciprocals of these intercepts, known as the *Miller indices*, are generally used. Thus, in a cubic crystal, one of the cube faces lies in or parallel to the plane of two principal axes and cuts the third at a distance equal to the length of the cube side, as shown in Fig. 3·10b. The intercepts of this plane are therefore ∞ on the x axes, 1 on the y axes, and ∞ on the z axes, and the Miller indices are $1/\infty$, $\frac{1}{1}$, $1/\infty$, or 010. This is usually written (010) in referring to the specific plane in question, or {010} in referring generally to planes of this form, and [010] when a direction perpendicular to the plane is indicated, as shown in Fig. 3·10c. The other cube faces have indices (100) and (001), as shown in Fig. 3·10d. Similarly, the diagonals that bisect the cube faces, as shown in Fig. 3·10e, are (110), (101), and (011) planes.

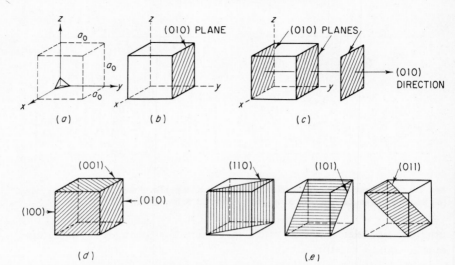

Fig. 3·10   Miller indices for simple planes and directions. (a) Cartesian coordinates; (b) the (010) plane of cubic crystals; (c) the 010 direction and the (010) family; (d) indices of other cube faces; (e) the (110) family of planes.

Two examples of more complex planes will be given. In Fig. 3·11a the plane makes intercepts of 1, 1, and $\frac{1}{2}$, and the indices are 1, 1, 2, thus the plane is the (112). The plane in Fig. 3·11b makes intercepts of $\frac{1}{3}$, $\frac{1}{2}$, and 1 and is therefore the (321).

3·19   Deformation of metals

If a metal is stressed very slightly, a temporary deformation takes place, permitted presumably by an elastic displacement of the atoms in the space lattices. Removal of the stress results in a gradual return of the object to its original dimensions. The conditions of such an experiment are expressed

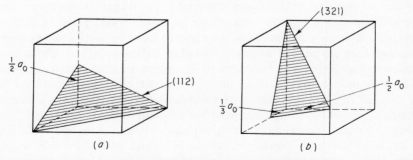

Fig. 3·11   Examples of complex planes in terms of Miller indices. (a) The (112) plane; (b) the (321) plane.

adequately by Hooke's law, given in Eq. (3·4), which says in the case of tension that strain e is proportional to the applied stress S.

$$e = \frac{S}{E} \tag{3·4}$$

where E = modulus of elasticity (Young's modulus)

S = stress, psi

e = strain, in./in.

Thus, service stresses in a metallic structure below the elastic limit (limit of the region where Hooke's-law behavior is followed) will usually produce deformation in accordance with the modulus of elasticity E of the metal and is entirely independent of its tensile strength and plasticity. If the metal is stressed beyond the elastic range, it may never return to its original form. Plastic deformation, or flow, of the metal under force has taken place.

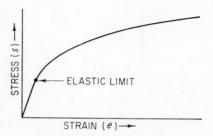

Fig. 3·12 Stress-strain curve.

The plastic flow of metals is perhaps their most important behavior, the one that distinguishes them most sharply from other construction materials. Some of the most important metallurgical operations known as *shaping* are in fact essentially examples of plastic deformation. For example, the stamping of automobile parts, spinning of aluminum pans, rolling of boiler plate, rails, and I beams, drawing of wire, extrusion of pipe, and forging of shafts are all processes involving the plastic deformation of metals. The changes produced in the properties of the metal so deformed are also important. We shall first consider the deformation of single crystals or grains.

If a pure single crystal of one of the cubic metals, such as copper or aluminum, is subjected to a gradually increasing stress, the elongation follows a curve of the type shown in Fig. 3·12. In the region near the origin, for low stresses, we see the linear elastic behavior as described by Hooke's law. As soon as the stress exceeds a certain value, which is to some extent dependent on the sensitivity of the measuring instrument, part of the elongation is permanent. The most common mechanism of deformation of this type is the process known as *slip*, which is characterized by the displacement of one part of the crystal relative to another along particular crystallographic planes. These planes may be detected by the presence of stepwise discontinuities on the surface of the specimen that are known as *slip bands*.

Plastic deformation of a cylindrical single crystal is shown schematically in Fig. 3·13a. The movement is usually along definite, predictable planes, similar to the motion in a distorted pack of cards. If a pack of cards were placed horizontally and a small force placed on the top card in a direction parallel to plane of the cards, the pack would distort about as pictured in Fig. 3.13*b*. The planes of a crystal are analogous to the planes on each of the playing cards.

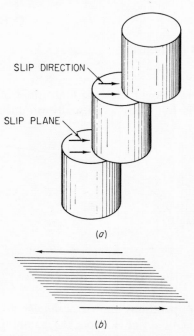

(*a*)

(*b*)

Fig. 3·13 Schematic representation of slip in a single crystal. (*a*) Slip on planes of a cylindrical single crystal; (*b*) distortion of cards by forces parallel to the planes of the cards.

3·20 Slip

The crystallographic planes on which movement takes place are called slip or glide planes. In fcc metals like copper and aluminum, the slip plane is usually the (111) plane. This plane has the greatest density and the largest spacing between planes. In other words, slip on this plane is easiest because the resistance to slip caused by parallel planes is a minimum. In hcp metals such as zinc, the slip plane is usually the basal plane, i.e., the close-packed plane. In bcc metals, there are generally several equally probable slip planes. For example, in alpha iron, slip can occur on the (110), (112), and (123) planes, all planes of high atomic density. (The student is advised to draw these planes and prove to himself that they are of similar density.) In deformations at higher temperatures, additional planes can also become active.

The *direction of slip* in a crystallographic plane is not the direction of greatest stress but a direction which coincides with the most closely packed rows of atoms in a lattice. This situation is represented schematically in Fig. 3·14. Note that slip may take place in any or all of the three directions shown. In fcc metals, the close-packed plane is the (111) plane, and the slip direction is the 101. The close-packed plane for hcp metals is the basal plane, and the slip direction is in a close-packed direction, as shown in

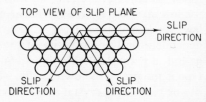

Fig. 3·14 Top view of slip plane, showing possible slip directions in a close-packed plane.

Fig. 3·14. In bcc metals with several possible slip planes, the 111 direction, which lies in each of the slip planes, is the slip direction.

The law governing the relation between slip and applied stress is a very simple one. For any given specimen, slip occurs along a given crystallographic plane when the component of shearing stress in that plane reaches a critical value. Although a crystal may be compressed or stretched, plastic deformation actually takes place by shear. Figure 3·15 represents the shear responsible for slip. Both the normal and shear stresses are shown in Fig. 3·15. It should be noted that the normal stress on the slip plane does not

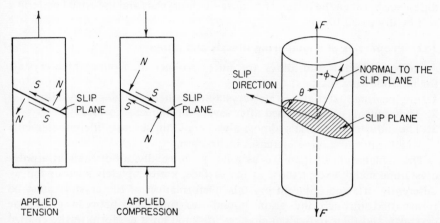

Fig. 3·15 Normal (N) and shear (S) stresses acting on the slip plane in tension and compression.

Fig. 3·16 Single crystal and angular relationships between the slip plane, slip direction, and applied force.

play a role in determining slip in crystals. When the shear stress reaches a critical value, slip begins. The magnitude of this stress can be calculated from the applied load. If we apply the tensile force F to the single crystal rod of Fig. 3·16, the *resolved shear stress* acting on any particular plane in a given direction can be calculated. The necessary information is the orientation of the plane and the direction of slip with respect to the axis of loading. If slip takes place on the plane whose normal makes an angle Φ with axis of the specimen, its area is given by,

$$A_{\text{slip plane}} = \frac{A}{\cos \Phi}$$

where A is the cross-sectional area of the specimen. The applied stress acting on the slip plane is therefore,

$$S_{\text{slip plane}} = \frac{F \cos \Phi}{A}$$

When this stress is resolved in the slip direction, the applied shear stress is

$$T_{\text{slip plane in slip direction}} = \frac{F \cos \Phi \cos \theta}{A} \tag{3·5}$$

The critical resolved shear stress is the lowest value of T that will just start slip. This value of $T_{critical}$ is dependent on composition, temperature, and the prior mechanical history of the specimen. For single crystals of pure metals, the values of critical resolved shear stress are usually between 50 and 500 psi. This relatively low value of $T_{critical}$ may cause some surprise when it is considered that most industrial metals and alloys have tensile strengths between 20,000 and 400,000 psi. The reasons for this very large difference between the behavior of pure single crystals and industrial materials will be discussed briefly.

3·21   Properties of engineering metals and alloys

Industrial metals and alloys are much stronger than pure single crystals for the following reasons:

1. Engineering materials are polycrystalline rather than single crystals.
2. These metals are often used after severe mechanical working.
3. The industrial materials almost always contain several different elements which can cause large increases in strength.

The explanation of item 1 is as follows: Normally, all crystals in a polycrystalline metal, except those at the surface, are completely surrounded by differently oriented adjacent crystals. Deformation of the crystals must be transmitted through the grain boundaries (interface between grains or crystals), and their individual changes in shape must conform to the change in shape of their neighbors. Immediately, we see the strengthening effect of polycrystallinity, since the deformation in any one grain is now restricted by surrounding grains of different orientation. The stress-strain curves for a single crystal and for a polycrystalline sample of the same metal are shown in Fig. 3·17. In general, the more crystallites there are in a unit volume (finer grain size), the harder and stronger the material will be.

Item 2 necessitates an examination of the changes in properties accompanying plastic deformation. That the properties of a metal should change appreciably when it is plastically deformed is perhaps surprising at first thought. That these changes, particularly in mechanical properties of the metal, should be such as to raise appreciably the strength of the metal is even more surprising. We shall start the examination of this phenomenon by returning to the stress-strain diagram.

In Fig. 3·18, the stress-strain diagram for a typical engineering polycrystalline metal is plotted. If we were to stress the material to point A and then release the load, the strain would not return to 0 but would remain at point B. The distance from the origin to point B denotes the amount of permanent deformation. However, if we were to reload the metal, no further plastic deformation would take place until the stress exceeded that given by point A. The virgin metal had previously begun plastic deformation after the elastic limit. Thus, the plastic deformation responsible for causing the permanent deformation of amount *OB* has caused an increase in elastic limit from point *E* to *A*. This phenomenon of hardening by plastic deformation is called *strain hardening*.

A few examples of the order of magnitude and importance of strain hardening will be mentioned. The strength of iron may rise from 25,000 up to 125,000 psi after very large amounts of deformation. A well-known example of strain hardening is the bending of a wire. The more times the wire is bent, the harder the wire, until the point is reached where the wire will no longer bend but fractures instead. This simple experiment explains

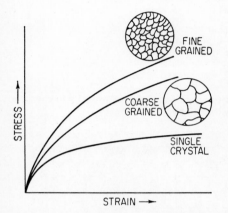

Fig. 3·17 Effect of grain size on stress-strain curves.

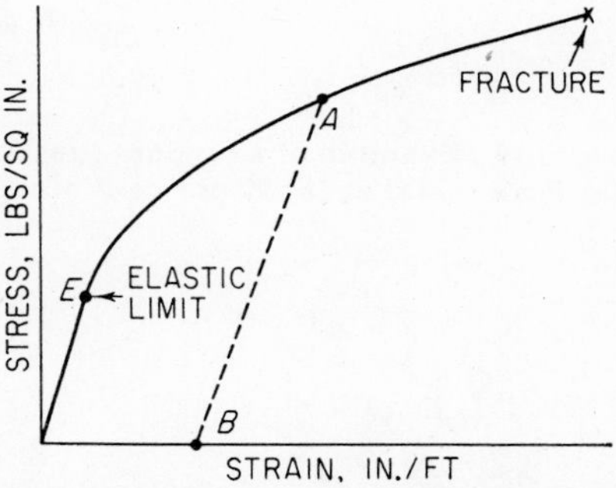

Fig. 3·18 Schematic representation of permanent deformation and strain hardening.

why copper wires for electric power transmission are hardened and strengthened by being drawn through a die. Their elastic limit is raised, and hence longer spans between poles can be used without danger of the wires breaking in a storm.

Item 3, hardening due to foreign elements, will be discussed briefly. A more detailed exposition on the theory and properties of alloys will be presented in the next chapter.

In the first examination of alloy systems, we find that there are two types of alloys: homogeneous and heterogeneous. A homogeneous alloy consists of a single phase (crystal structure), and the composition of the alloy is the same from grain to grain. In this case, the homogeneous alloy may be visualized in the following manner: If the pure metal consists of *x* atoms and the alloying element of *y* atoms, then the homogeneous alloy can be represented by Fig. 3·19. The *y* atoms are being substituted for *x* atoms, and therefore this type of alloy is called a *substitutional solid solution*.

A heterogeneous alloy consists of two or more phases, each phase having a

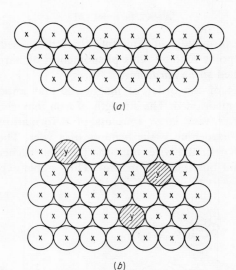

Fig. 3·19 Example of a random substitutional solid solution of metal y in metal x. (a) Pure metal x; (b) 90 per cent metal x and 10 per cent metal y.

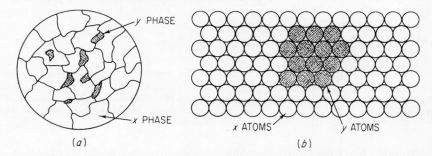

Fig. 3·20 Example of a heterogeneous alloy. (a) Microscopic scale; (b) atomic scale.

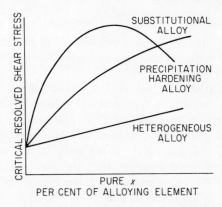

Fig. 3·21 Effect of alloy type and composition on critical resolved shear stress.

different composition. Figure 3·20 represents a heterogeneous alloy containing two phases from an atomic and a microscopic viewpoint. In the example of Fig. 3·20, one phase consists of pure x atoms and the other of pure y atoms.

In general, dissolved alloying elements, as demonstrated by the substitutional alloys, harden a given pure ,metal more than undissolved ones, so that it might be said that, for a given composition, an alloy is generally harder when in the homogeneous than in the heterogeneous form. There is a very important exception to this rule, namely, the case of *precipitation hardening*. This class of alloys will be discussed further in the following chapter. Figure 3·21 summarizes the effect of composition on the critical resolved shear stress of typical engineering alloys.

3·22 Annealing of deformed metals

As a metal is deformed, we have seen that it hardens and may eventually fracture. Now let us turn to the problem of resoftening a deformed metal. In the deformed condition, the atoms of a metal are in a state of high internal energy. This high internal energy can be released by heating the metal. The metal atoms return to their equilibrium positions, and the original properties of the metal are gradually recovered. The shape as given the object by the deformation (also called cold work), however, remains unchanged during heating. The heating of a cold-worked metal to or above the temperature at which equilibrium begins to return is called annealing.

The effects of heating on cold-worked metal may be divided into three general stages:
1. Recovery
2. Recrystallization
3. Grain growth
Let us heat-treat a severely cold-worked metal at various temperatures and note the changes that occur. Figure 3·22 shows the results of such an experiment.

As the temperature of a cold-worked metal is gradually raised, the first change that takes place is the relief of internal stress (recovery). The behavior of the electrical resistivity parallels that of internal stress. These properties recover without any accompanying change in the microstructure. The *recrystallization* reaction, which occurs at a higher temperature than recovery, is a phenomenon in which entirely new, *strain-free* grains are formed from the deformed metal. The physical properties are all returned to their original values at the end of this range. Recrystallization starts by the formation of tiny new grains (usually at grain boundaries); it continues as the small grains grow and new grains are nucleated until all the distorted, elongated grains have been replaced. Further heating to higher temperatures will cause the strain-free grains to grow at the expense of other strain-free grains; this phenomenon is *grain growth*.

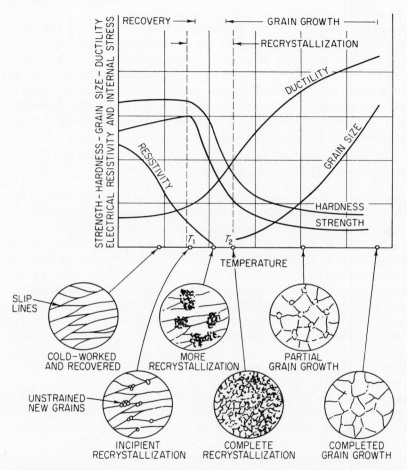

Fig. 3·22 Effect of annealing (constant time) on properties and microstructure of cold-worked metal.

3·23 Summary

By this time, the student should have a basic concept of the nature and characteristics of metals. He must now let these ideas "jell" in his mind so that each part can be related to the central concept. In order to facilitate this review, a short summary of the basic ideas will be presented:

   1. In general, metal atoms contain one or two electrons in their outermost electronic shell.

2. In the formation of a solid, these electrons become part of an "electron gas" which is responsible for the characteristic properties of metals.
3. In the formation of chemical compounds, the metal atoms tend to lose their outer electrons and become positively charged ions.
4. The most important properties of metals are strength and plasticity.
5. Other important properties are corrosion resistance, electrical and thermal conductivity, ferromagnetism, etc.
6. In metals, the atoms are arranged in a three-dimensional lattice structure which makes up the individual grains or crystals of a metal.
7. The three most common structures found in metals are the face-centered cubic, body-centered cubic, and hexagonal close-packed.
8. In general, a metal consists of many small regions (grains) in which the atoms are all arranged in the same three-dimensional array. These grains are separated from one another by the grain boundary, which is a region of misfit between grains.
9. Planes and directions in crystals are described in terms of a system of Miller indices which is related to the unit cell of a metal.
10. When a metal is stressed elastically, it returns to its original dimensions when the stress is relieved. If the metal is stressed beyond the elastic range, plastic deformation takes place and the metal may never return to its original form.
11. The mechanism of plastic deformation is known as slip, which is characterized by the displacement of one part of the crystal relative to another along particular crystallographic planes.
12. Slip takes place on planes of greatest density in a direction coinciding with the most closely packed rows of atoms on the slip plane.
13. Slip occurs along a given crystallographic plane when the component of shearing stress reaches a critical value.
14. Engineering metals and alloys are much stronger than pure single crystal because:
    a. Engineering materials are polycrystalline.
    b. Metals are often used after they have been strain-hardened.
    c. Alloying elements are added to harden the pure metal.
15. A deformed metal may be resoftened by heating. The effects of heating or annealing on cold-worked metals is generally divided into three stages:
    a. Recovery.
    b. Recrystallization.
    c. Grain growth.

PROBLEMS

In the following multiple-choice problems, the reader should select the correct answers (one or more).

I. The periodic table classifies the elements according to (*a*) melting point; (*b*) atomic weight; (*c*) metals and nonmetals; (*d*) atomic number.

2. The element titanium (atomic number 22) has (*a*) a completely filled *d* shell; (*b*) a partially filled *d* shell; (*c*) a completely filled 3*p* shell; (*d*) a partially filled 3*s* shell.

3. Ionization refers to (*a*) a metal atom losing an electron and becoming a negatively charged ion; (*b*) a metal atom losing an electron and becoming a positively charged ion; (*c*) a nonmetal atom gaining an electron and becoming a negatively charged ion; (*d*) the characteristic that individual atoms tend to arrange themselves so that the outer electronic shells are completely filled.

4. The differences between metals and nonmetals are due to (*a*) the number of electrons in the filled shells; (*b*) the number of electrons in the outer shells; (*c*) the atomic weight; (*d*) the density.

5. Some of the more important properties of metals are (*a*) combination of strength and plasticity; (*b*) combination of strength and brittleness; (*c*) corrosion resistance; (*d*) thermal and electrical conductivity.

6. Special properties of metals include (*a*) diamagnetism; (*b*) high specific heat; (*c*) high melting point; (*d*) thermoelectric power.

7. Metals crystallize in (*a*) the hcp structure; (*b*) the simple cubic structure; (*c*) fcc and bcc structures; (*d*) orthorhombic and tetragonal structures.

8. Examples of crystalline polymorphy are (*a*) the transformation of alpha to beta zirconium at 863°C; (*b*) the transformation of beta to gamma uranium at 760°C; (*c*) the melting of iron; (*d*) the cooling of iron from 910 to 900°C.

9. Several close-packed structures can be considered to be made up of close-packed atoms; for example, (*a*) an ABCABC packing gives the bcc structure; (*b*) an ABAB packing gives an hcp structure; (*c*) an ABCABC packing gives an fcc structure; (*d*) an ABCABABCAB packing gives the orthorhombic structure.

10. The plastic behavior of metals is characterized by (*a*) a process known as slip; (*b*) Young's modulus; (*c*) permanent deformation; (*d*) elastic behavior.

II. Slip usually refers to (*a*) atomic movements along planes of lowest density; (*b*) atomic movements along particular planes in particular directions; (*c*) motion along the (110) plane in fcc metals; (*d*) motion along the (111) plane in fcc metals.

12. Engineering metals are much stronger than pure single crystals because (*a*) they are polycrystalline; (*b*) they are of greater density; (*c*) they are cooled rapidly from the liquid state; (*d*) they are usually alloyed.

4

THE SCIENCE OF METALLURGY

4·1 Introduction

The science of metallurgy covers the broad fields extending from the processing of mined deposits to the final step of using the metal for an engineering application. In particular, we shall divide this large field into the three topics listed below:

1. Process metallurgy

2. Physical metallurgy

3. Fabrication metallurgy

The present discussion will deal with the problems confronting the metallurgist once an ore deposit has been found and a concentrated ore supply is available. The field of process metallurgy is concerned with extracting the pure metals from the naturally occurring mineral deposits. Once a commercially pure metal has been produced, the job of the process metallurgist is essentially done. At this time, it is the function of a physical metallurgist to test and evaluate the metal and its alloys. This study is utilized by the fabrication metallurgist to produce useful metal products.

PART I. PROCESS METALLURGY

4·2 Mineral dressing

Metals are extracted from the earth's crust by sequences of operations and processes. First, the metallic minerals must be found in the earth's crust in deposits of sufficient size and metal concentration, and with other characteristics which will make extraction of the metal profitable. The metal is mined as an ore, or mineral mixture, which usually contains sizable proportions of waste minerals mixed and intergrown with the valuable metallic minerals. For most ores, the next step, carried out in mineral dressing plants, is to separate the valuable minerals. The rejection of waste minerals is generally accomplished by first crushing and grinding the ore to sever the minerals from each other and then concentrating the ore by such processes as *gravity concentration* (separating minerals of different densities), *froth flotation* (a concentrating method based on the adhesion to air of some particles from a pulp and the simultaneous adhesion of other particles to water), and *magnetic separation* (separation of ferromagnetic minerals from others). These mineral dressing operations generally do not involve changes in the physical or chemical identities of the minerals which are separated.

4·3   Liberation of metals

The next step in winning metal from ore is to liberate it from the mineral with which it is chemically combined. Metals occur as sulfides, carbonates, or silicates or as more complex chemical compounds. The metal may be extracted from such compounds by *leaching* (solution by strong acids) and precipitation, by *electrolysis* (an electrical method of separating elements) of solutions or fused salt mixtures, or by methods involving high temperatures. Some copper ores are leached with acid, the copper being precipitated from solution with iron scrap or shot. The copper may sometimes be pure enough to electroplate directly from such solutions. Aluminum is won by fusing the oxide (Al_2O_3) with sodium aluminum fluoride (cryolite, Na_3AlF_6) and electrolyzing the mixture.

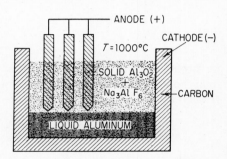

Fig. 4·1   Hall process for extracting aluminum by electrolysis.

This process of electrolysis was discovered by Hall and Heroult in 1886 and is the basis for the present aluminum industry. In the Hall process, the electrolysis is carried out in large iron pots with a thick carbon lining which acts as a cathode. A number of large graphite rods sticking down into the pot serve as the anode, as shown in Fig. 4·1. The graphite rods are first lowered until they touch the cathode and an arc is struck; powdered cryolite is then added and melted by the heat of the arc. When a sufficient liquid bath is obtained, aluminum oxide is added and the anodes drawn farther away from the cathode. The addition of the oxide raises the resistance of the liquid somewhat. The temperature of the bath is kept at about 1000°C, and since this is above the melting point of aluminum (660°C), the molten aluminum collects as a liquid in the bottom of the cell and is drawn off at intervals. Oxygen is liberated at the anode and gradually burns away the graphite. The cell reaction is

$$2Al_2O_3 = 4Al + 3O_2 \qquad\qquad (4·1)$$

Usually metals are more economically liberated from mineral concentrates by thermal (pyro-) methods. In pyrometallurgy, the metal is seldom obtained in one step from the ore concentrate. If the ore is a sulfide, it is commonly first roasted in air to form an oxide. Oxide ores, as well as roasted sulfides, are heated in vertical-shaft (blast furnaces) or hearth furnaces with a reducing agent such as coke or charcoal, which releases the metal by the following schematic reaction:

$$MO_2 + C = CO_2 + M \qquad (4.2)$$

Limestone is added as a flux to form a light (low-density) fluid slag by combining with the extraneous minerals, the slag separating from the liquid metal by gravity. The liquid metal produced is cast into "pigs" of convenient size or is transferred in the molten state to an adjacent refinery.

4.4   Refining of metals

Refining is defined as the operation of purifying crude metals. Copper, zinc, and sometimes gold, silver, and lead are refined by electrolytic methods;

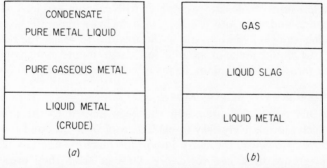

Fig. 4·2   Metal-refining process. (a) Distillation; (b) pyrometallurgical refining.

zinc, mercury, and magnesium, by *distillation*; and pig iron, and often copper and lead, by pyrometallurgical operations. In distillation, shown schematically in Fig. 4·2a, volatile metals are removed from the crude metal as a gas and are condensed as a high-purity liquid. The phases present in pyrometallurgical refining are given in Fig. 4·2b. In this operation, the general procedure is to melt the crude metal, add reagents, and then control the temperature and chemical conditions so that the impurities are removed from the liquid metal and collected in a separate phase.

As an example of pyrometallurgical refining, we shall examine the refining or iron as carried out in the open-hearth furnace. The starting product is principally "pig iron" from the blast furnace, and the final product is generally a *low-carbon steel*. *Steel* is an alloy of iron plus small but carefully controlled amounts of carbon, manganese, phosphorus, sulfur, silicon, and small amounts of other metals. The composition limits for a few grades of steel are given in Table 4·1. These limits are determined by the Society of Automotive Engineers (SAE) and the American Iron and Steel Institute (AISI). The last two numbers of the SAE designation give the approximate carbon content of the steel.

The sequence of operations in the open-hearth furnace start by charging with scrap steel, limestone, and iron ore. When this charge is melted under an open flame, the molten pig iron is added. Most of the impurities such as

Table 4·1   Composition limits of standard low-carbon steels

SAE number	Carbon	Manganese	Phosphorus maximum	Sulfur maximum
1010	0.08–0.13	0.30–0.60	0.040	0.050
1030	0.28–0.34	0.60–0.90	0.040	0.050
1060	0.55–0.65	0.60–0.90	0.040	0.050

silicon, manganese, phosphorus, and carbon of the original charge are oxidized. The nonvolatile oxides combine with calcium and iron oxides (from the limestone and ore) into a slag, which forms a separate layer over the metal. After sufficient time to allow the mixture to reach the desired composition, the liquid metal is poured into large ingots, which will then be fabricated into one of the many forms of commercial steel.

4·5   Unusual processing problems

Up to this point, the discussion has been limited to the processing of metals which are relatively easy to extract from their ores and fabricate into useful forms. However, the more important reactor materials fall into the rare-metal group; these are not easily extracted from their ores, and their fabrication is considerably more difficult than that of the common metals. These characteristics have necessitated the application of many processing techniques which are considerably different from the more classical ones mentioned previously in this chapter. In fact, in some instances it has been necessary to develop a special technique before certain of the rare metals could be prepared in anything approaching a pure form or a desired shape.

Several examples of some of the unusual techniques which have been applied to the production of rare metals will be discussed. The discussion will be limited to the obviously important reactor materials such as uranium, thorium, beryllium, zirconium, etc.

4·6   Metal replacement

Metal replacement is carried out by reacting an *active metal* with the oxides or halogen compounds of the rare metals so that the active metal replaces the rare metal in the oxide or halogen compound. An *active metal* is one which has a greater affinity for oxygen or the halogen than the rare metal. The usual active metals used in metal replacement are calcium, sodium, and magnesium. The metal replacement reaction can be written schematically as:

$$M_{active} + M_{rare}O \rightarrow M_{active}O + M_{rare} \qquad (4\cdot3)$$

Titanium and zirconium are made by the Kroll process, which reduces the tetrachlorides of these metals with molten magnesium. Thorium and vanadium are made by the reduction of their oxides with calcium, and uranium is obtained by the calcium reduction of uranium fluoride.

4·7 Halide decomposition

Owing to the rather low melting and boiling points which characterize metal halides, in general, these compounds will decompose thermally to yield the metal. Thus, halide decomposition can be described by the following equation:

$$M_{rare}H_{halide} + \text{heat} \rightarrow M_{rare} + H_{halide} \qquad (4\cdot4)$$

The iodides of titanium, hafnium, zirconium, vanadium, thorium, and uranium can be decomposed on a heated surface, such as an electrically heated tungsten wire, in an evacuated container to form the corresponding massive metals in a state of very high purity. Commercially, this iodide process is to be regarded more as a purification step than as a primary production process, but several of the pure metals have been prepared initially if not solely by it.

4·8 Inert-atmosphere arc melting

The great affinity of many of the rare metals to oxygen, nitrogen, hydrogen, and carbon, plus the fact that many of them are obtained from the initial reduction stage in the form of finely divided powers of great surface area and reactivity, requires the use of special techniques for the formation of massive metal forms. Tungsten, molybdenum, tantalum, and columbium have long been fabricated by powder metallurgy operations involving the *sintering* of a bar (which had been compacted under high pressure) in an atmosphere of hydrogen. Sintering of metal powders refers to the bonding of adjacent surfaces of particles in a mass of powder or a metal compact by heating. Molybdenum and several other metals, such as titanium and zirconium, are now being melted in arc furnaces to form ingots. A compacted bar of the metal powder or sponge (product of halide decomposition) is fed as a consumable electrode into an arc furnace maintained under an inert atmosphere of helium or argon. The molten metal is collected in a water-cooled copper crucible which serves as the second electrical terminal.

In essence, the container for the molten metal is a solidified shell of the metal itself, and the usual pickup of impurities from a ceramic crucible is eliminated. Figure 4·3 shows the general layout during arc melting. A nonconsumable tungsten electrode can also be used to strike the arc for melting, in which case the metal is fed directly into the molten pool on the lower electrode; however, the trend is toward the consumable electrode type of furnace, to avoid contamination of the metal by the nonconsumable electrode material. Alloys can also be prepared in these furnaces by feeding the alloying elements in appropriate fashion to the molten pool.

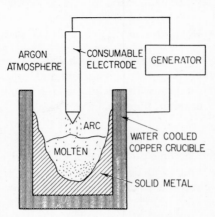

Fig. 4·3   Consumable electrode arc melting.

4·9   Typical metal-processing flow sheets

As a means of unifying the important principles of metal processing in the form of a brief summary, two complete flow sheets will be presented. The flow sheets for the extraction of iron and uranium are shown diagrammatically in Figs. 4·4 and 4·5, respectively. These flow sheets are greatly condensed, to show only the essential steps and paths of materials flow.

The flow sheet for the extraction of iron ore is relatively straightforward and needs no further discussion. The uranium extraction flow sheet is an order of magnitude more difficult than that of iron. The first step consists in making a physical separation of magnetite and light gangue after initial crushing in the ball mill. In the next few steps, the ore is treated with a mineral acid to bring the uranium into solution. Then the uranium is converted to a soluble complex carbonate, which allows removal of insoluble iron, aluminum, and cobalt bases. The uranium is recovered as sodium uranate ($Na_2U_2O_7$) or ammonium uranate [$(NH_4)_2U_2O_7$] after several intermediate steps which allow removal of additional metal impurities. In the final step, the uranium oxide is obtained as a precipitate from the ammonium uranate.

The production of uranium metal from the oxide can be carried out by the following methods:

1. Reduction of uranium oxides with carbon
2. Reduction of uranium oxides with aluminum, calcium, or magnesium
3. Reduction of uranium halides with alkali metals or alkaline-earth metals
4. Fused salt electrolysis of uranium halides
5. Thermal decomposition of uranium halides

From the available data on production of pure metallic uranium, the general conclusion can be reached that the methods using purified oxides or halides with purified alkali or alkaline-earth metal reductants will produce the purest uranium metal.

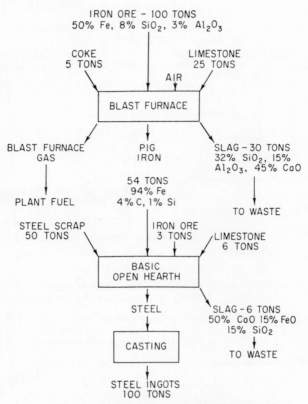

Fig. 4·4 Extraction of iron from iron ore.

PART 2. PHYSICAL METALLURGY

4·10 Alloys

A study of physical metallurgy is the prime requisite for understanding how to use metals intelligently. Most of the subject matter discussed in Chap. 3 falls under the heading of physical metallurgy. We shall start the following discussion by developing a detailed picture of *alloying*.

Alloys are of great importance in industry, in fact, far more so, quantitatively considered, than are the pure metals. The steels, the brasses and bronzes, aluminum alloys, bearing metals, and cast iron recall at once the importance of this class. Pure metals are desired for certain purposes, such

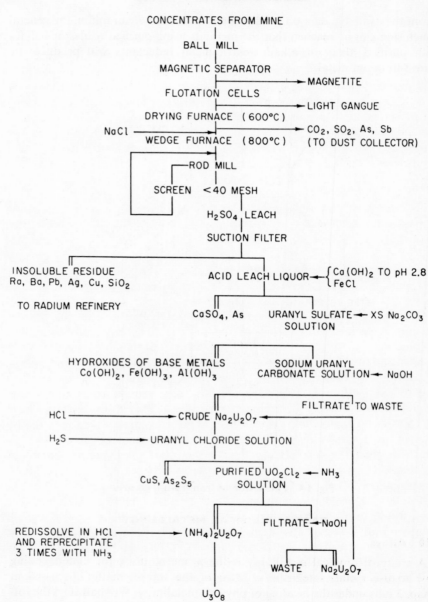

Fig. 4·5   Flow sheet for production of U_3O_8 from Canadian pitchblende.

as copper and aluminum for electrical conduction and aluminum, titanium, and tin for corrosion resistance. But far greater quantities of these metals are used in the form of alloys with other metals and even with small proportions of nonmetals, such as carbon, nitrogen, sulfur, and oxygen. It is the purpose of this section to develop the concepts of the *phase diagram*, which gives a graphic description of the effects of alloying on the structure of the component metals. When this structure is known as a function of composition, many of the properties, both physical and chemical, can be predicted, because the structure of an alloy is responsible for its particular properties.

4·11   Phase diagrams

The study of phase diagrams is by far one of the most important single topics in the field of physical metallurgy. Before we discuss the more theoretical details of the significance of phase diagrams, several simple examples will be described. The simplest example of alloying occurs when two metals are completely *immiscible* in both the liquid and the solid state; i.e., neither of the two metals is soluble in the other. An alloy system which demonstrates this effect is the aluminum-lead system.

The first step in making alloys consists in preparing a series of mixtures containing various compositions of aluminum and lead from pure aluminum to pure lead. If we then heated these mixtures until they became completely molten and then cool, we should have made a series of aluminum-lead alloys. If, during the cooling of these alloys from the liquid state, we were to record temperature as a function of time, a number of interesting results would become evident. Pure aluminum would solidify from the liquid state at 660°C, and the cooling curve would appear as curve *A* in Fig. 4·6. The molten aluminum cools rapidly until the freezing temperature of 660°C is reached; then the temperature remains constant until all the aluminum solidifies. Once complete solidification has taken place, the temperature, once again, falls off rapidly with time.

Cooling of an alloy containing 75 per cent aluminum and 25 per cent lead would produce the cooling curve *B* in Fig. 4·6. Because of complete immiscibility between aluminum and lead, all the aluminum freezes at 660°C, but the lead remains liquid until 327°C. Similar reasoning explains curve *C* in Fig. 4.6. The last cooling curve in Fig. 4·6 is that of pure lead.

The phase diagram in Fig. 4·7 summarizes the information obtainable from a whole series of cooling curves for aluminum-lead alloys. The base line, or abscissa, of the phase diagram shows the compositions of all possible alloys from 0 per cent of one metal and 100 per cent of the other metal at one end of the line to 100 per cent of the one metal and 0 per cent of the other at the other end of the line, for any two metals or nonmetallic components. It is thus possible to represent any binary alloy by a point on this

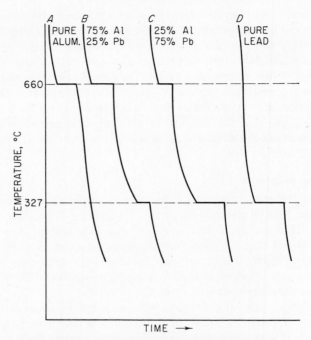

Fig. 4·6 Cooling curves of aluminum-lead alloys.

scale intermediate between its two extremes, as, for instance, 86 per cent of one component and 14 per cent of the others, as shown at x in Fig. 4·7. The vertical scale of the diagram represents the temperature range. Any point within the confines of the diagram will therefore represent a particular alloy composition at a particular temperature.

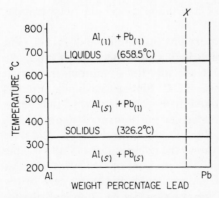

Fig. 4·7 Aluminum-lead phase diagram (simplified slightly).

Let us continue with a more detailed examination of Fig. 4·7. Above 660°C, the alloys consist of two liquid phases of pure aluminum and pure lead. If these alloys were kept molten, the lead would settle to the bottom and the aluminum would rise to the top because of the large difference in density between aluminum and lead and because the liquids are immiscible. Between 660 and 327°C, alloys of aluminum and lead consist of two phases, one solid and the other liquid. Below 327°C, all alloys would be solid, consisting of two solid phases, pure aluminum and pure lead. The line running across the diagram at 660°C is called the *liquidus*, and it indicates the temperature at which freezing begins for each alloy. Above the liquidus, the alloy is entirely liquid. At 327°C, solidification is complete for all aluminum-lead alloys, and this line is called the *solidus*. Below the solidus, all phases are solid.

4·12 The bismuth-cadmium phase diagram

Other, rather simple, examples of phase diagrams are those in which the liquid phases are *miscible*, i.e., soluble in one another, but where there is

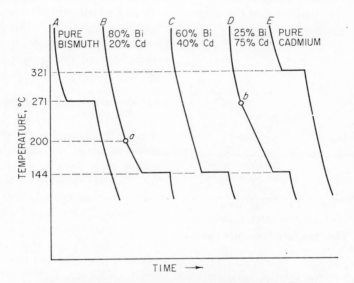

Fig. 4·8 Cooling curves for bismuth-cadmium alloys.

complete immiscibility in the solid phases. These characteristics are displayed in the bismuth-cadmium system. The cooling curves for a series of bismuth-cadmium alloys are shown in Fig. 4·8. Curves *A* and *E*, representing pure bismuth and cadmium, respectively, show the same general behavior as the curves for the pure metals lead and aluminum. However, curves *B*, *C*, and *D* illustrate new phenomena. Both curves *B* and *D* have an abrupt change in slope of the cooling curve at the temperatures indicated by letters *a* and *b*. If we could examine these alloys during cooling from the molten state, we would find that a solid phase just starts to appear at the change in slope. As cooling continues, the quantity of the solid phase increases. Since the solid phases are completely insoluble in one another, the solid phase forming at point *a* in curve *B* is pure bismuth and at point *b* in curve *D* it is pure cadmium. The complete explanation of the cooling curves should become clearer after one further point is discussed.

All cooling curves of the bismuth-cadmium alloys, except those of the pure components, show a hold at 144°C. This hold in the cooling curve occurs at the *eutectic* temperature. At the eutectic temperature, all the remaining liquid completely solidifies according to the eutectic reaction, which may be written as

$$\text{Liquid} \underset{\text{heating}}{\overset{\text{cooling}}{\rightleftharpoons}} A_{\text{crystals}} + B_{\text{crystals}} \tag{4.5}$$

Since the cooling curve C shows no break until the eutectic temperature is reached, this signifies that the 60 per cent Bi and 40 per cent Cd alloy is the eutectic composition. Figure 4.9 gives the phase diagram which summarizes all the cooling curves described.

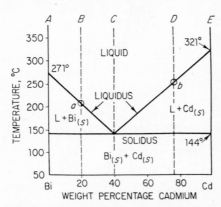

Fig. 4.9 The bismuth-cadmium phase diagram.

4.13   General rules of phase diagrams

Let us first examine the physical interpretation of the shape of the liquidus curve of Fig. 4.9. For an alloy system in which the components are completely soluble in one another in the liquid state but completely insoluble in the crystalline state, *Raoult's law* states that the freezing point of a pure substance will be lowered by the addition of a second substance. For an alloy of Raoult's type, Fig. 4.10 shows a freezing-point curve (liquidus curve) beginning at the freezing point of the pure metal and dropping steadily with each addition of the second substance. Above the liquidus, the alloy is entirely liquid; immediately below, the melt has begun to form solid crystals of pure A. Therefore, the region below the liquidus curve is a *two-phase field* containing one liquid and one crystalline phase. The quantity and composition of each of these two phases, one liquid and one crystalline, for any alloy at any temperature may be found by using two simple rules.

First, let us choose a composition x, as shown in Fig. 4.10, and examine this alloy composition in the two-phase field at temperature T_1. Pure component A is soluble in the liquid phase at T_1 up to the composition y. This result can be obtained by the first rule.

Rule 1. To determine the composition of either the liquid or the solid when both are present, one need only draw a horizontal exploring line at the chosen temperature. The intersections of this line with the two boundaries of the two-phase field give (by projection on the abscissa) the compositions of each of the two phases existing at that temperature. For example, at T_1, the horizontal line intersects the liquidus at y, and that of the solid is pure A. As the temperature is lowered further, more A crystals appear and the composition of the liquid follows down along the liquidus.

Rule 2. Not only is the composition of each of the two phases that can be present at any temperature shown by this diagram, but also the relative quantity of each of these two phases present in any chosen original composition is given. To determine the quantity of each phase, it is necessary to construct a vertical line at the point on the composition of the mixture as line xpq in Fig. 4·10. The point p where this vertical line intersects the chosen temperature horizontal line oy may be considered the fulcrum of a lever system. The relative lengths of the lever arms op and py multiplied by the amounts of the phases present must balance. For example, in an alloy containing 90 per cent A and 10 per cent B, as shown at q in Fig. 4·10, the amount of A present in the alloy at temperature T_1 is

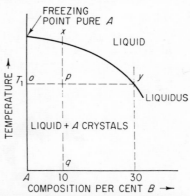

Fig. 4·10 Schematic phase diagram showing the liquids as a function of composition.

$$\frac{py}{oy} \times 100 = {}^{20}\!/_{30} \times 100 = 66.67 \text{ per cent } A$$

and the amount of liquid of composition y present is

$$\frac{op}{oy} \times 100 = {}^{10}\!/_{30} \times 100 = 33.33 \text{ per cent liquid}$$

Thus, the two rules give both the composition of each phase and the relative quantity of each of the two phases present in any binary system. Now let us return to the discussion of the phase diagram of Fig. 4·9.

We shall examine, in detail, the cooling of an alloy made up of 80 per cent Bi and 20 per cent Cd. References will be made to cooling curve B of Fig. 4·8 and to the phase diagram of Fig. 4·9. When the temperature of the liquid alloy reaches the liquidus at point a, a sharp break in the cooling curve is observed. The reason for the discontinuity is that heat is liberated during solidification. As the temperature falls below the liquidus, the amount of solid Bi increases and the composition of the liquid moves down along the liquidus. For example, at a temperature of 175°C, the composition of the liquid is 70 per cent Bi and 30 per cent Cd, and the relative amounts of solid and liquid are 33.33 per cent Bi and 66.67 per cent liquid. At 144°C, the liquid composition reaches the eutectic point of 60 per cent Bi and 40

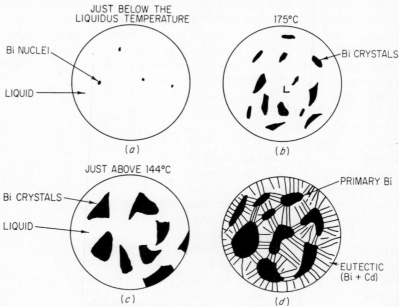

Fig. 4·11   Photomicrographs of an 80 per cent bismuth, 20 per cent cadmium alloy during cooling. (a) Minute amount of Bi crystals; (b) 33 per cent Bi solid, 67 per cent liquid, 30 per cent Cd, and 70 per cent Bi; (c) 50 per cent Bi solid and 50 per cent liquid of composition 40 per cent Cd and 60 per cent Bi; (d) 50 per cent Bi solid and 50 per cent eutectic (40 per cent Cd and 60 per cent Bi mixture of phases).

per cent Cd. At this temperature, 50 per cent of the total alloy has solidified as Bi crystals. Just below 144°C, the alloy is completely solidified as the remaining liquid is solidified by the eutectic reaction

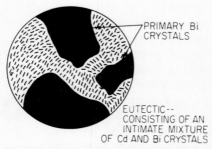

Fig. 4·12   Schematic representation of the details of the eutectic mixture of cadmium and bismuth crystals.

$$\text{Liquid}_{\text{solution Bi and Cd}} \rightarrow \text{Bi}_{\text{crystals}}$$
$$+ \text{Cd}_{\text{crystals}}$$

A series of photomicrographs taken during the cooling procedure described above would appear as shown in Fig. 4·11. These figures should be self-explanatory. A more detailed picture of the eutectic is given in Fig. 4·12. The eutectic structure consists of an intimate mixture of Bi and Cd crystals. Figure 4·13

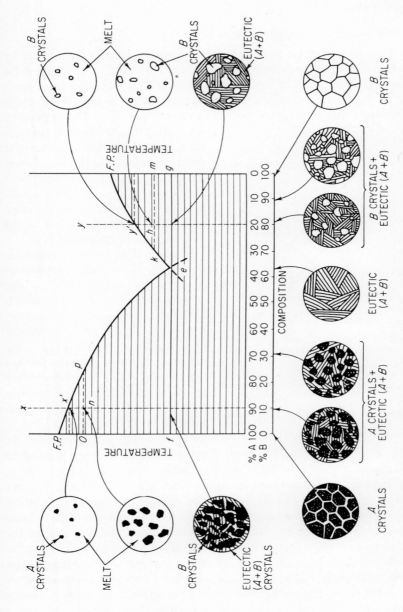

Fig. 4·13   The microstructure of alloys during solidification and at room temperature.

graphically and pictorially represents the phase diagram and structure of alloys of Bi and Cd.

4·14   The copper-nickel phase diagram

The third and last phase diagram to be discussed in detail is that formed by two substances which are completely soluble in both the liquid and the solid state. A typical diagram of this kind is given in Fig. 4·14, that of copper-nickel. In this case of complete solubility, the freezing points of the alloys fall constantly as the proportion of Cu in them increases. The upper line connecting the freezing point of pure Cu and Ni is the liquidus curve, at which solidification begins during the cooling of each alloy. The lower line is the solidus, at which solidification is complete. The enclosed area is a two-phase field of one liquid and one crystalline phase.

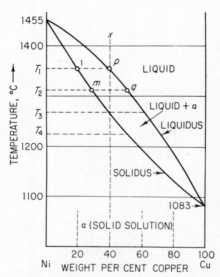

Fig. 4·14   The copper-nickel phase diagram.

If we cool an alloy of composition x (Fig. 4·14), the first crystals of a appear at temperature T_1. The composition of the solid a which forms at T_1 is given by the intersection of the solidus and a horizontal line at point 1. As the alloy cools, the composition of the solid follows the solidus curve. At temperature T_2, the liquid composition is given by point q and the solid composition by point m. For the solid to change its composition during cooling, for example, from 1 at the start of crystallization at T_1 to T_2 near the end of crystallization, *diffusion* within the solid phase must take place. Diffusion is defined as a movement of atoms within any substance. The net movement is usually in the direction from regions of high concentration toward regions of low concentration, in order to achieve homogeneity. Since the phase diagram describes the equilibrium condition, diffusion allows the solid phase a to alter its composition with decreasing temperature. However, if the alloy is cooled too rapidly to allow the a phase to homogenize, the resulting solid phase will consist of cores or rings of varying composition, as shown in Fig. 4·15. The equilibrium cooling permits the formation of homogeneous *equiaxed* grains of a, while the nonequilibrium cooling gives rise to a *cored* structure within equiaxed grains. An equiaxed grain possesses equal diameters in all directions, as shown at temperature T_4 in Fig. 4·15. A cored

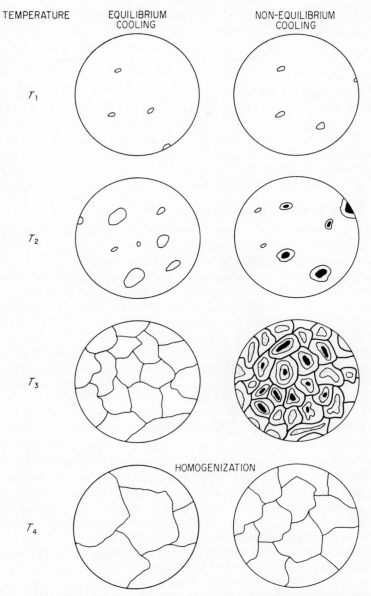

Fig. 4·15 Microstructure of solid-solution alloys after equilibrium and non-equilibrium cooling.

structure shows a concentration gradient consisting roughly of spherical shells, each of a slightly different composition, somewhat analogous to the structure of an onion. If the cored alloy is then heated to a temperature just below T_4 and held for some time, diffusion will gradually homogenize the *a* grains.

Before passing to the more complicated types of phase diagrams, the student is urged to try his hand at solving the first few problems given at the end of this chapter. Over 50 per cent of the problems of metallurgists or materials engineers require as a start a thorough examination of the phase diagrams of the alloys involved. Therefore, the student must master this subject matter in order to be in a position to appreciate many of the problems and much of the information given in the later sections of this book.

4·15   Complex binary phase diagrams

The two extreme types of alloys, those completely insoluble in the crystalline state and those completely soluble, have now been examined. However, most alloys are intermediate between these two and show partial solubility in the solid state. An understanding of this type of phase diagram requires no new principles, but merely slight modifications of the extreme types already examined. The lead-tin diagram of Fig. 4·16 shows partial solubility in the solid state. The lines *lm* and *no* are merely curved slightly and moved closer to their corresponding liquidus curves, and the two two-phase fields, resembling in form the more dilute regions of the two-phase field of Fig. 4·14, are obtained. The crystals that precipitate from these liquid alloys are not pure *A* or pure *B* but always contain a little of the other component in solid solution.

4·16   The precipitation-hardening reaction

The behavior of the liquidus curves and the eutectic in Fig. 4·16 are the same as described for the case of complete insolubility of solid phases. However, below the temperature of the eutectic line of Fig. 4·16 an interesting and very important phenomenon takes place. As the temperature falls, there is a decrease of the solubility of *B* in crystals of α and a decrease in the solubility of *A* in crystals of β. This is the phenomenon responsible for the precipitation- and age-hardening alloys.

The microstructure of alloys between points *m* and *o* would appear the same as those pictured in Fig. 4·13. Alloys to the left of *m* and to the right of *o* would show the precipitation phenomenon. A series of microstructures of an alloy of composition *x* at temperatures T_1, T_2, T_3, and T_4 is shown in Fig. 4·17. At T_1, the alloy is two-phase as α crystals are forming from the melt. The liquid composition moves down along the liquidus as the temperature drops, and the solid α composition follows the solidus. At T_2, the alloy is a single-phase solid solution, consisting of *a* grains of composition

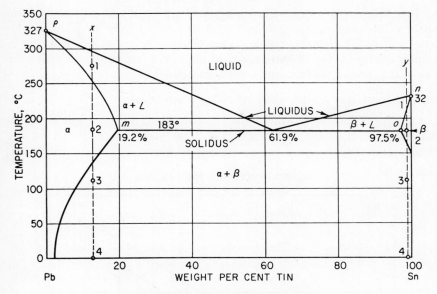

Fig. 4·16   The lead tin phase diagram.

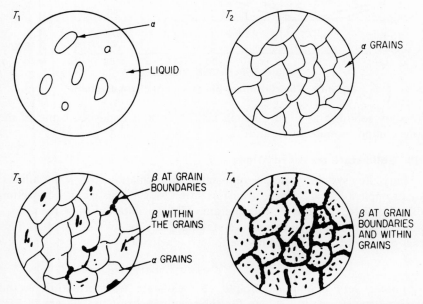

Fig. 4·17   Microstructure of a precipitation-hardening alloy during cooling from the liquid state.

x. Just below the two-phase boundary at T_3, β begins to precipitate from the α solid solution. The precipitate forms preferentially on grain boundaries and in the crystallographic planes of the α crystals. This behavior is exactly analogous for alloys to the right of point o, except for the difference that α is precipitating from the β solid solution.

4·17 Completely soluble alloys showing a minimum

Figure 4·18, below, shows a binary system which demonstrates complete solubility in both the liquid and the solid state but differs from the usual cigar-shaped figure by possessing a minimum. At the minimum,

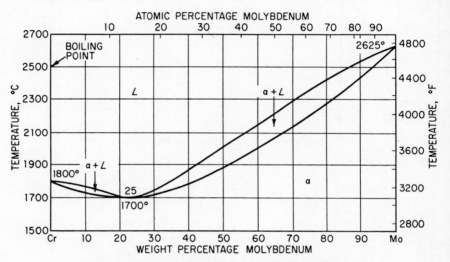

Fig. 4·18 The chromium-molybdenum phase diagram.

the alloy solidifies as a pure metal, i.e., all crystals deposited being of the same composition as the liquid.

4·18 Solid-state transformations

Many alloy systems undergo solid-state transformations below the solidus. An example of a solid-state transformation was discussed in connection with Fig. 4·16, i.e., the Pb-Sn diagram. The precipitation of β from α_{solid} solution was a solid-state reaction taking place below the solidus. This reaction may be represented by the following equation:

$$\text{Supersaturated } \alpha_{solid} \rightarrow \beta_{solid} + \text{equilibrium } \alpha_{solid} \qquad (4\cdot6)$$

Another very common example of solid-state transformation is the *eutectoid* transformation shown for the uranium-chromium system in Fig.

4·19. Below the solidus, two *eutectoid* transformations take place as the alloys are cooled to room temperature. The eutectoid transformation is

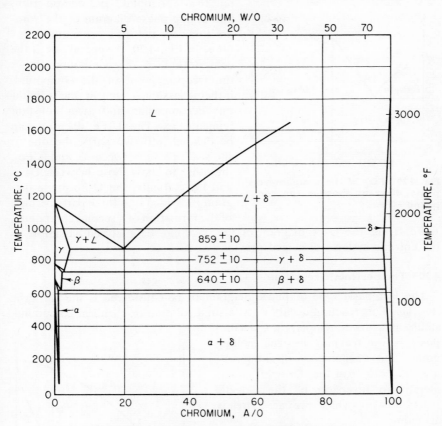

Fig. 4·19 The uranium-chromium phase diagram.

similar to the eutectic reaction except that the liquid phase is replaced by a solid phase as shown in the following equations:

Entectic reaction:

$$\text{Liquid} \rightarrow \alpha_{\text{solid}} + \beta_{\text{solid}} \qquad (4\cdot7)$$

Eutectoid reaction:

$$\gamma_{\text{solid}} \rightarrow \alpha_{\text{solid}} + \beta_{\text{solid}} \qquad (4\cdot8)$$

4·19 Compound formation in binary diagrams

The combination of two kinds of atoms into an alloy results not only in liquid or solid solutions but sometimes in a chemical combination between the atoms to form a compound, such as Fe_3C in steel, $CuAl_2$ in duraluminum, UFe_2, or UBi in a reactor fuel slurry. Such a chemical combination is shown

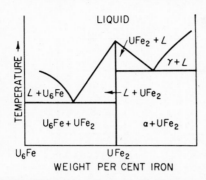

on the horizontal, or composition, line of the phase diagram at the composition of the new compound, as UFe_2 in Fig. 4·20. A vertical line at the composition of the compound shows the freezing point of the compound, it being assumed that it melts without decomposing and at a constant temperature. The U-Fe diagram can be divided into two parts, the one to show all alloys between U and UFe_2, the other to show those between UFe_2

Fig. 4·20   Part of the uranium-iron phase diagram showing the compound UFe_2.

and Fe. Actually, the U-Fe phase diagram is divided into three parts because of the formation of a second compound, U_6Fe. Thus, the U-Fe diagram of Fig. 4·20 can be analyzed as simply as the eutectic systems discussed previously.

4·20   The peritectic reaction

The last basic type of phase diagram to be considered is the *peritectic reaction*. This reaction occurs when a solid solution or chemical compound, stable at a lower temperature, decomposes before reaching its true melting point. This phenomenon, it is interesting to note, can be thought of as merely an inversion of the eutectic reaction with which we are already acquainted.

Upon heating an unstable alloy, let us say compound A_xB_y of Fig. 4·21, to its decomposition temperature, the A_xB_y crystals would yield a liquid 1, which is different in composition from A_xB_y, in this case richer in A. The formation of this liquid phase leaves crystals that are no longer A_xB_y but that perhaps are almost pure B with a small solubility for A, as shown in the diagram. The peritectic reaction is characterized by decomposition of a homogeneous phase such as A_xB_y at a certain temperature, upon heating, into two other phases, external in composition to the original A_xB_y. The peritectic reaction

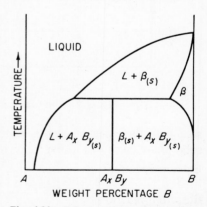

Fig. 4·21   Schematic representation of a phase diagram containing a peritectic.

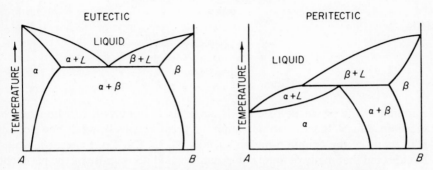

Fig. 4·22   Comparison between eutectic and peritectic reactions.

might be more clearly understood by comparing it with the eutectic reaction. The symbolic equations for the peritectic and eutectic reactions are given in Table 4·2 for the case of cooling and heating. These reactions are also pictured schematically in Fig. 4·22.

The peritectic reaction occurs in many important alloys, such as copper-zinc, iron-nitrogen, and many others. The copper-zinc diagram of Fig. 4·23 shows five peritectic reactions and one eutectoid reaction. A thorough under-

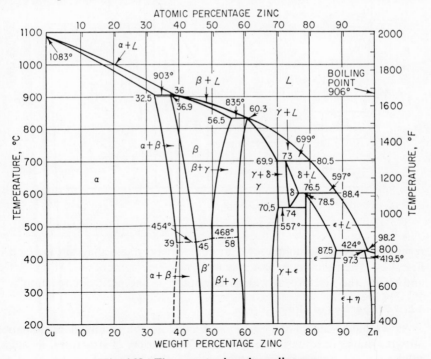

Fig. 4·23   The copper-zinc phase diagram.

Table 4·2

	Peritectic	*Eutectic*

Peritectic   *Eutectic*

COOLING

Liquid + $\beta \rightarrow \alpha$ Liquid $\rightarrow \alpha + \beta$

HEATING

$\alpha \rightarrow$ Liquid + β $\alpha + \beta \rightarrow$ Liquid

standing of the peritectic reaction is particularly important for the study of uranium alloys. Since uranium tends to form relatively weak compounds with many of the common metals such as Al, Bi, Cu, Co, Fe, etc., it would be expected that many of these compounds would decompose by the peritectic

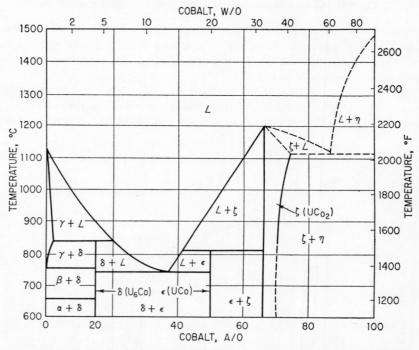

Fig. 4·24   The uranium–cobalt phase diagram (not completely determined).

reaction. A typical example of a uranium metal phase diagram is given in Fig. 4·24. The uranium–cobalt system forms three compounds, U_6Co, UCo, and UCo_2, the first two of which decompose by the peritectic reaction.

4·21   Properties of alloys

As we have seen in the previous discussion, the solidified combination of two metals can result in the formation of a variety of structures, including solid solutions, intermetallic compounds, and mixtures of phases as in the eutectic structure. The properties of alloys are derived from these structures and do not depend upon any linear relation as a function of alloy composition. To illustrate this point, several properties of the Cu-Ni and Pb-Sb systems will be discussed.

4·22   Solid-solution alloys

The Cu-Ni system, as previously given in Fig. 4·14, forms complete solid solutions. The meaning of the term solid solution will be explained with the

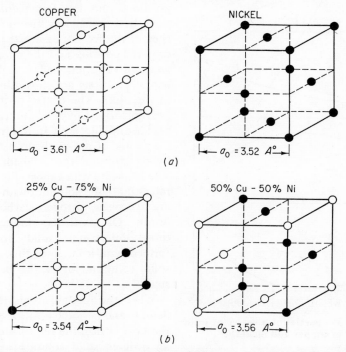

COPPER

NICKEL

$a_0 = 3.61$ A°

$a_0 = 3.52$ A°

(a)

25% Cu – 75% Ni

50% Cu – 50% Ni

$a_0 = 3.54$ A°

$a_0 = 3.56$ A°

(b)

Fig. 4·25   Schematic representation of a substitutional solid solution of copper and nickel. (a) Crystal structures of pure copper and nickel; (b) solid solutions of copper and nickel.

aid of Fig. 4·25. The crystal structures of Cu and Ni are both fcc. The solid solutions of Cu and Ni are also fcc and are pictured for two compositions, 25 per cent and 50 per cent Cu (Fig. 4·25b). The Cu and Ni atoms are positioned randomly with respect to one another on the fcc lattice. The lattice parameters of the solid solutions are between the parameters of pure Cu and pure Ni but do not follow a linear relationship as a function of composition, as shown in Fig. 4·26. This type of solid solution is called a *substitutional solid solution* because metal *B* atoms are just substituted for metal

A atoms as *B* is added to *A*. Two other important types of solid solutions, the *interstitial solid solution* and the *ordered solid solution*, are pictured in Figs. 4·27 and 4·28.

An interstitial solid solution will form, generally, when the solute atoms are of a very much smaller atomic diameter than the solvent atoms. In Fig. 4·27, the solute atoms fit into the spaces between the solvent atoms in the lattice. In the bcc lattice, the interstitial sites are the face-centered positions and the *tetrahedral positions*. In a similar way, the interstitial sites in the fcc lattice are the body-centered position and the *tetrahedral positions*. The tetrahedral and octahedral positions are shown in detail in Fig. 4·27. The center of a tetrahedron formed by four atoms is the tetrahedral site, and in like manner the center of an octahedron formed by six atoms is the octahedral site. The elements which most often form interstitial solid solutions with metals are hydrogen, nitrogen, carbon, and boron. Of particular interest are those interstitial solid solutions formed with iron, since this type of solid solution forms the basis for steels.

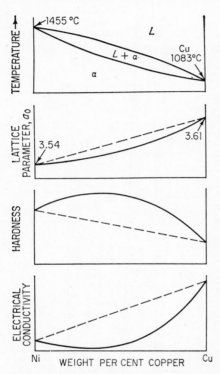

Fig. 4·26   Properties of a series of solid solutions of copper and nickel.

When the foreign atoms take up definite and periodic positions within the host lattice, the requirements for an ordered solid solution are met. Figure 4·28 illustrates an ordered solid solution of equal composition of *A* and *B* atoms. This type of ordered solid solution is formed in copper-gold alloys, and the ordered phase is written as CuAu.

Let us return to the detailed discussion of Cu-Ni alloys. As mentioned previously, the lattice parameter of solid solutions of Cu and Ni does not vary linearly with composition but deviates noticeably, as shown in Fig. 4·26. The hardness (resistance to indentation, often used as a qualitative measure of strength) also does not behave linearly with composition but rises steadily from both pure metals to a maximum at 50 atomic per cent. It might be somewhat surprising that the addition of the softer Cu to the harder Ni should increase the hardness of nickel. However, the student should realize that the problem is, not the question of the hardness of the individual components, as in the case of a two-phase alloy, but that of the effect of the solute atoms upon the solvent lattice and the nature of the

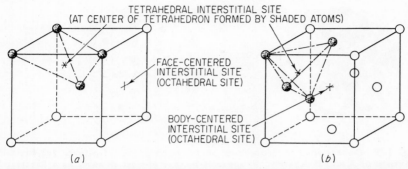

TETRAHEDRAL INTERSTITIAL SITE
(AT CENTER OF TETRAHEDRON FORMED BY SHADED ATOMS)

FACE-CENTERED
INTERSTITIAL SITE
(OCTAHEDRAL SITE)

BODY-CENTERED
INTERSTITIAL SITE
(OCTAHEDRAL SITE)

(a)   (b)

Fig. 4·27   Tetrahedral and octahedral interstitial sites in the bcc and fcc unit cells. (a) Interstitial sites in the bcc unit cell; (b) interstitial sites in the fcc unit cell.

lattice forces operative, owing to the interaction of the different atomic species. In other words, when Cu is added atom by atom into the Ni lattice, the interactions between the Cu and Ni atoms determine the resultant properties. Another example of this type of phenomenon is shown by the behavior of the electrical conductivity as a function of the Cu concentration. At 50 atomic per cent, the conductivity is lowest, suggesting that the interactions between Cu and Ni might be at a maximum. This maximum interaction between Cu and Ni atoms may also explain the hardness maximum at 50 atomic per cent.

As the theory of the metallic state is not sufficiently advanced to predict quantitatively the interactions between

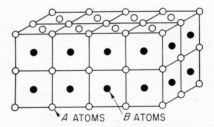

A ATOMS   B ATOMS

Fig. 4·28   Ordered substitutional solid solution (copper-gold type).

atoms which might affect such things as lattice distortion and the distribution of electrons, we must be satisfied with a qualitative description. For our purposes it is sufficient to know that these interactions do exist and that they cause solid-solution alloy properties to vary nonlinearly with composition.

4·23   The lead-antimony alloys

The properties of an alloy series of two metals nearly insoluble in the solid state, as in the case of Pb-Sb alloys, vary almost linearly with the proportionate weights of each of the components. The linear relationship is dependent to a great extent upon an even distribution of the constituents of the mixed series. If, however, one constituent completely surrounds the other, the properties of the resultant alloy will depend almost entirely upon the continuous metal matrix. Figure 4·29 illustrates this point.

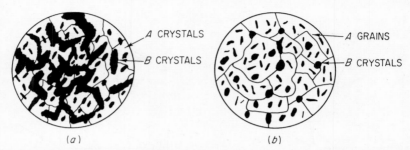

Fig. 4·29 Dependence of properties on microstructure of alloys. (a) Intimate mixture of A and B crystals. Properties behave almost linearly as a function of composition. (b) Metal A forms continuous matrix. Properties depend almost entirely on metal A.

Two properties of the Pb-Sb alloys are summarized in Fig. 4·30. The hardness of Pb-Sb alloys varies almost linearly between points *b* and *o* in the figure. On the basis of the previous discussion of solid-solution alloys, it

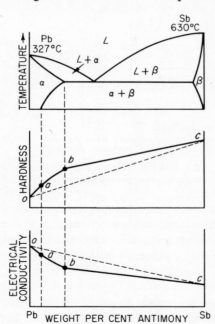

is no surprise that the line between points *o* and *a* does not lie on the straight line indicated by the dashed line. However, one might expect a straight line from point *a* to point *b*. The reason for the continuation of the curved line between points *o* and *a* has to do with the mechanism by which the *a* solid solution adjusts its composition during cooling from 252°C to room temperature. This mechanism is called *precipitation hardening* and will be discussed in the following section on Heat-treatment. The electrical conductivity behaves as expected, with almost linear relationship of conductivity vs. composition in the two-phase region (between points *b* and *c*).

Fig. 4·30 Properties of lead-antimony alloys.

4·24   Heat-treatment

Heat-treatment may be defined as a combination of heating and cooling operations, timed and applied to a metal or alloy in the solid state in a way that will produce the desired properties. Heat-treatment can be used to secure the following changes in properties:

1. To harden and strengthen metals
2. To relieve internal stresses, as in castings
3. To anneal objects previously deformed cold
4. To change the distribution or size of particles in a multiphase alloy

Items 2 and 3 listed above have already been discussed in the previous chapter in the section on Annealing of Deformed Metals. The student is advised to reread the aforementioned section before proceeding. Items 1 and 4 form the basis for the following discussion.

4·25 Nonequilibrium cooling

Up to this point in this discussion of the nature of metal alloys, we have examined only the equilibrium condition. For example, phase diagrams,

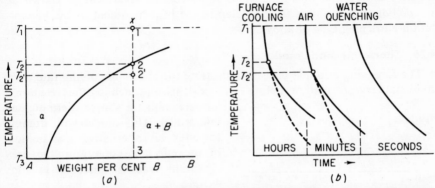

Fig. 4·31 Cooling of precipitation-hardenable alloys. (a) The solid solubility of B in solid solution as a function of temperature; (b) cooling curves for various cooling techniques.

which form the basis for much of this section, give the composition of the phases present under equilibrium conditions, i.e., by extremely slow heating or cooling. In metallurgical practice, however, these requirements for equilibrium are almost never fulfilled. Alloys are heated and cooled many orders of magnitude faster than permissible to allow equilibrium reactions to take place.

As an example of a nonequilibrium situation, let us examine Fig. 4·31, which illustrates the effect of cooling rate on a typical solid-state transformation such as the precipitation of β from supersaturated α solid solution. The schematic phase diagram of Fig. 4·31a shows the composition x of the alloy under consideration and the temperature T_1 at the start of cooling, T_2, the temperature at which transformation would start under equilibrium conditions, and T_3, room temperature. The cooling curve for a furnace-cooled sample (cooling rate about 0.5°C per minute) initially at T_1 is shown in Fig. 4·31*b*. This curve drops smoothly until T_2, where a break and change in slope are observed, the precipitation reaction being thus identified. The dotted lines show the extrapolations of the upper part of the cooling curves. By removing a sample at T_1 from a furnace and allowing it to cool in air, a cooling rate of about 500°C per minute could be attained. Under these conditions, the cooling curve would be described by the second curve of Fig. 4·31*b*. The break in the cooling curve appears at T_2', which is somewhat below T_2, indicating that the precipitation reaction has been suppressed by air cooling. Water-quenching the sample after removal from the furnace could result in a quenching rate as high as 40,000°C per minute. Such a quench could completely suppress the solid-state precipitation reaction and produce the third cooling curve of Fig. 4·31*b*. The complete suppression of the solid-state reaction means that the α phase of composition x has been retained at room temperature. This rapid-cooling treatment is known as *solution quenching* and is the first step in the heat-treatment of the precipitation-hardening type of alloys.

4·26   Thermodynamic model

The foregoing example served to illustrate the effects of cooling rate on a solid-state transformation. Before extending this approach to other reactions, let us examine a simple thermodynamic model. This model will attempt to offer an alternative explanation of the effect of cooling on the precipitation reaction pictured in Fig. 4·31*a*.

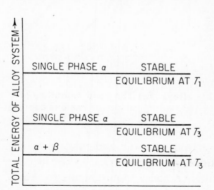

Fig. 4·32   Schematic illustration of the total energy of a precipitation-hardenable alloy system.

If we were to plot an energy-level diagram for the alloy of composition x at T_1 and T_3 for the furnace-cooled and water-quenched samples, it would appear as in Fig. 4·32. The total energy of the alloy system is plotted vertically as a function of temperature and heat-treatment. The total energy includes the thermal energy and the energy of interaction of A and B atoms.

At T_1, for the single-phase alloy, the thermal energy is high, and the interaction energy is 0; that is, A and B atoms are in a state of equilibrium as part of the α phase at temperature T_1. At T_3, two conditions of equilibrium exist. For the furnace-cooled sample at T_3, the thermal energy is low, and once again the interaction energy is 0 because of equilibrium. However, for the water-quenched sample at T_3, the thermal energy is low, but the inter-action energy is high, because at T_3 the α solid solution of composition x is not the stable-equilibrium phase. The α phase of composition x retained at room temperature is in metastable equilibrium.

Figure 4·33a demonstrates pictorially the concept of metastability. The long board is stable when balanced along an edge, as long as there are no forces tending to tip it over. However, the board could lower its center of gravity or, in terms of this analogy, its potential or internal energy by falling on its side. It would then be in its lowest stable state. The force required to

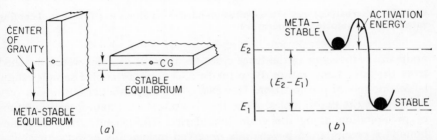

Fig. 4·33 Schematic illustrations of the concepts of metastability and activation energy. (a) Schematic model showing difference between metastable and stable equilibrium; (b) schematic model of a transition from one energy state to another.

upset the metastable equilibrium is called the *activation energy*. Once the activation energy has been attained, the system lowers its energy spontaneously. Figure 4·33b shows two energy states E_2 and E_1 separated by an energy barrier, which is just the activation energy. In this example, state E_2 is metastable and would spontaneously change to the stable state E_1 after the required activation energy is supplied.

4·27 Heat-treatment of precipitation-hardening alloys

Let us return to the discussion of the heat-treatment of precipitation-hardening alloys. As stated previously, the effect of the water-quenching treatment is to retain the α phase at room temperature. This supersaturated phase is metastable and thus could be made to transform to the stable two-phase alloy by supplying the necessary activation energy. In this particular example, the activation energy is supplied by heating the supersaturated solid solution to temperatures between T_3 and T_2. This type of heat-treatment is called *aging*.

The general effects of aging on the microstructure and properties of a precipitation-hardening alloy will now be discussed. First, we shall examine the microstructure of the alloys cooled by three methods outlined in Fig. 4·31b: furnace cooling, air cooling, and water quenching. Figure 4·34 shows the typical microstructures obtained in precipitation-hardening alloys. The

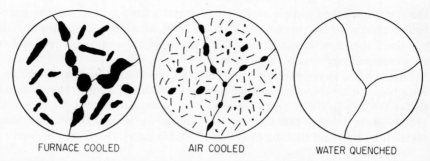

FURNACE COOLED   AIR COOLED   WATER QUENCHED

Fig. 4·34 Microstructures of precipitation-hardening alloys as a function of cooling rate from the solution temperature.

grain size is the same in the three cases because the samples were quenched from the same temperature. Both the furnace- and air-cooled samples show large amounts of precipitation. However, the precipitate particle size is very different in the two samples. Also, the two samples contain a large amount of precipitation along the grain boundaries. The phenomenon of grain-boundary precipitation is generally observed, principally because precipitation can start earliest in the rather jumbled lattice at the grain boundaries. The water-quenched sample remains single-phase, since the high-temperature condition of complete solid solution is retained at room temperature.

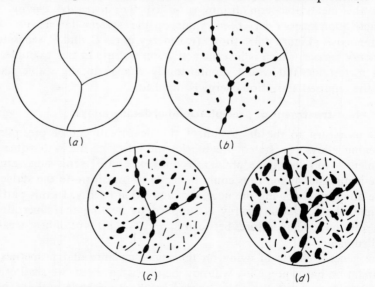

Fig. 4·35 Effects of aging on the microstructure of a precipitation-hardening alloy. (a) As water quenched; (b) start of aging; (c) maximum mechanical properties; (d) overaged condition.

Starting with a water-quenched sample, let us trace the effects of aging on the microstructure. As shown in Fig. 4·35, the first evidence of precipitation appears at the grain boundaries and to a lesser degree within the grains. As aging continues, the initial precipitate grows and the *nucleation* of additional particles causes the appearance of new particles principally within the grains. This stage is generally associated with maximum mechanical properties, as will be described below. *Nucleation* describes the process of forming small β particles from the supersaturated solid solution. The last stage of aging shows the coarsening of the precipitates and is characterized by a decrease in mechanical properties. The term *overaged* is used to describe this condition.

4·28 Atomic model of precipitation

As an aid in forming a mental picture of the mechanics of precipitation, an atomic model is shown in Fig. 4·36. A solid solution can be considered as a random mixture of *A* and *B* atoms. During the early stages of aging, the *B* atoms tend to cluster and form nuclei, from which the precipitate

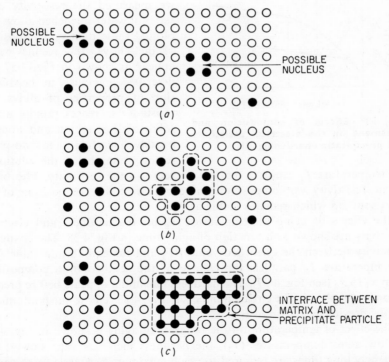

Fig. 4·36 Atomic model of the formation of a precipitate particle. (*a*) Random solid solution; (*b*) formation of nucleus; (*c*) formation of precipitate particle.

particles are created. During the birth of the precipitate particle, it breaks away from the parent matrix by creating an interface between it and the matrix and then adjusts its lattice parameter to the requirements of the B atoms. The new interface and lattice configuration cause some straining in the adjacent matrix lattice. This straining of the matrix is partly responsible for the hardening which takes place during precipitation.

4·29   Property changes during aging

The changes in microstructure during the course of aging are reflected in the physical properties. Two properties, hardness and electrical resistivity,

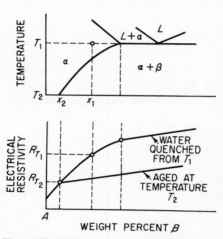

Fig. 4·37 Effects of composition and treatment on the electrical resistivity of a precipitation-hardening alloy.

will be examined as a function of aging time and temperature. The hardness values represent the macroscopic strength of the alloy. Electrical-resistivity values give information which can be related to the composition of the supersaturated solid solution. For example, if we measured the resistivity of a series of water-quenched alloys and plotted these data vs. composition of the alloys, the results would appear as shown in Fig. 4·37. As described previously in the discussion of properties of alloys, the resistivity increases rapidly in the solid-solution region and approximately linearly in the two-phase region. Aging below the solutionizing temperature T_1 causes a marked decrease in the resistivity. The break in the resistivity vs. composition curve occurs at the composition of the solid solution which remains stable at T_2.

The effects of aging at temperature T_2 on the hardness and electrical resistivity are shown as a function of aging time in Fig. 4·38. The change in resistivity between the quenched value RT_1 and the equilibrium value RT_2, at temperature T_2 parallels the change of the solid-solution composition from x_1 to x_2 (see Fig. 4·37). The hardness maximum appears before precipitation is completed. The circled numbers in Fig. 4·38 refer to the photomicrographs previously shown in Fig. 4·35.

The effect of temperature on the progress of aging is shown in Fig. 4·39. At low aging temperatures (T_4), the hardness increases very slowly, and extremely long times are required to reach maximum hardness. As the aging temperature is increased, the time required to approach maximum hardness decreases. At very high temperatures, precipitation occurs at extremely fast rates, and the maximum hardness attained is considerably below that achieved at lower aging temperatures. In practice, aging temperatures between T_2 and T_3 are used, and the alloy is aged to maximum hardness for most applications.

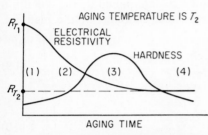

Fig. 4·38 Effects of aging at temperature T_2 on the hardness and electrical resistivity

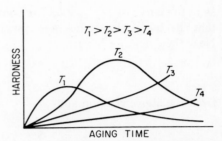

Fig. 4·39 Effects of aging temperature on hardness of a precipitation-hardening alloy.

4·30 Imperfections in crystal lattices

No survey of physical metallurgy would be complete without a discussion of the effect of lattice imperfections on the properties of metals. Lattice imperfections are introduced during solidification of crystals from the melt. Research on solidification of metals has shown that in the absence of impurities or solid particles lattice dislocations are necessary for the growth of crystals from the melt. Only through very carefully controlled techniques can crystals be prepared with relatively few imperfections, and then these crystals are only the size of whiskers. When a near-perfect whisker is tested mechanically, critical shear stresses several orders of magnitude higher than those found for normal metal crystals are observed. The reasons for this effect will be explained.

In Chap. 3, the values of critical resolved shear stress for pure metals were given to be between 50 and 500 psi. These are empirical values obtained by mechanical testing of single crystals. Let us examine a simple method of estimating the theoretical shear strength of a crystal. Consider the shearing of two rows of atoms past each other in a homogeneously strained crystal, as in Fig. 4·40a. Let the spacing between the rows be a and that between atoms along the slip direction be b, and suppose that the shear displacement of the upper row over the lower one is x when the shear stress acting on them is S. Now, consider a relation between the shearing force and the shear displacement. Clearly, the force is 0 at positions such as A and B, since these are the normal lattice sites, and also at positions halfway between them, by symmetry. Each atom is attracted toward its nearest lattice site, as defined by the atoms of the other row, so that the shearing force must be a periodic function of x, with period b. Assume that it is sinusoidal,

$$S = k \sin \frac{2\pi x}{b}$$

as in the curve of Fig. 4·40b. The constant k is determined by the condition that the initial slope must agree with the shear modulus of the material. Near the origin we have

$$S = \frac{k 2\pi x}{b} \tag{4·9}$$

from the above relation and

$$S = \gamma G = \frac{x}{a} G \tag{4·10}$$

from Hooke's law, in which G is the shear modulus and γ is the shear strain; since Eqs. (4·9) and (4·10) must be the same, we have

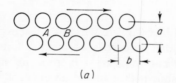

(a)

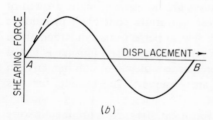

(b)

Fig. 4·40 Calculation of theoretical shearing force. (a) Shearing of two rows of atoms past each other; (b) sinusoidal shearing force.

$$S = \frac{b}{a} \frac{G}{2\pi} \sin \frac{2\pi x}{b} \tag{4·11}$$

The critical shear stress, at which the lattice becomes mechanically unstable and slip should take place, is thus

$$S_{max} = \frac{b}{a} \frac{G}{2\pi} \tag{4·12}$$

Since a and b are almost equal, the theoretical shear strength is

$$S_{max} = \frac{G}{2\pi} \tag{4·13}$$

which is several orders of magnitude greater than the observed value for single crystals. For example, G equals about 15×10^6 psi for copper; therefore, the theoretical shear stress required for slip is 10^6 psi. This is approximately 10^4 times larger than the critical shear stress measured in copper single crystals.

How can we explain this large discrepancy between the experimental and theoretical values of critical shear stress? The answer was suggested independently by E. Orowan, G. I. Taylor, and M. Polanyi in 1934. They suggested that slip occurs by the movement of *dislocations*. A dislocation is

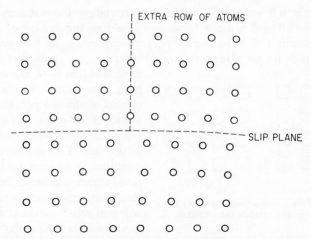

Fig. 4·41   A unit edge dislocation in a simple cubic lattice.

defined as the boundary between slipped and unslipped regions of a slip plane. A dislocation is shown schematically in Fig. 4·41. The edge dislocation given can be represented by an extra row of atoms which ends at the slip plane.

Let us examine the stresses involved in moving a dislocation. Figure 4·42 illustrates motion of a dislocation in several steps under the action of the applied shear stress. A dislocation in a crystal is highly mobile in the sense that, when it moves on the slip plane, the resistance that the lattice offers to its motion is small and may be overcome by a small applied stress, several orders of magnitude smaller than the shear modulus. A detailed treatment of the shear stress required to move a dislocation is beyond the scope of this text; however, one can see in a general way from the model of Fig. 4·42 why a dislocation is mobile.

Dislocation theory plays a major role in the field of physical metallurgy. Modern theories of the yield point, strain hardening, and precipitation hardening rely on dislocation theory for a theoretical basis.

Two other types of lattice imperfections will be described. They are *lattice vacancies* and *interstitial atoms*. These are referred to as point imperfections, whereas a dislocation is a line imperfection. Figure 4·43 illustrates these two kinds of point imperfections.

A lattice vacancy is simply the hole existing when a

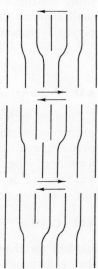

Fig. 4·42 Schematic representation of the motion of a dislocation under an applied shear stress.

lattice site is left vacant. The atoms surrounding the vacancy are some-what perturbed and cause a local stress field in the region of a vacancy. Metals contain a few vacancies per million atoms at low temperatures. Near the melting point, experimental measurements show that the vacancy concentration is of the order of 100 ppm. The molten state is believed to consist of about 3 per cent vacancies.

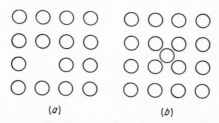

(a)   (b)

Fig. 4·43 Point imperfections. (a) Lattice vacancy; (b) lattice interstitial.

Interstitial atoms are atoms which are in the interstices rather than on lattice sites, This is different from the condition in an interstitial solid solution, which was described earlier in this chapter. In an interstitial solid solution, small atoms, such as hydrogen and carbon, fit under equilibrium conditions in the interstices of the metal lattice. Interstitial atoms, on the other hand, are atoms, trapped in interstitial sites, which would otherwise prefer to be at the normal lattice sites.

Both interstitials and vacancies play an important role in the theory of radiation damage. Fast neutrons colliding elastically with lattice atoms cause the formation of vacancies and interstitials. In the next chapter, we shall examine the details of radiation damage and in particular the part played by vacancies and interstitials.

PART 3. FABRICATION METALLURGY

4·31 Production processes

Starting with pure metals and alloys and a thorough knowledge of the properties of these materials, the fabrication metallurgist supervises the production of useful objects. The operations involved in fabricating metals include casting, powder metallurgy, forging, extrusion, drawing, rolling, welding, etc. In this part, we shall examine the principles of these operations and shall also discuss several practical design problems and their solutions.

4·32 Casting

Most metal products start out by being poured, or "cast," while liquid into a fireproof container, or "mold." If they are permitted to retain this shape in use or are reshaped by machining operations only, the objects are called castings. If they are to be subjected to subsequent deformation such as rolling, forging, etc., the original shapes are called, not castings, but *ingots*, *billets*, or *blanks*. However, regardless of whether the objects are castings or ingots, the basic requirements for each are the same: good density and strength properties, and the least possible waste in pouring and subsequent operations.

The first step in casting is the melting of the metal. Melting can be carried out in many different types of furnaces, the choice of furnace depending upon many factors. For the sake of brevity, the economic considerations of melting practice will not be discussed. Among the engineering factors affecting the choice of a furnace are:

1. Temperature required
2. Degree of control of temperature
3. Atmosphere (vacuum of inert)
4. Size of melt
5. Crucible and mold materials

Fuels or power sources for melting metals and the types of furnaces used are listed in Table 4·3. Examples of the metals melted by the various furnaces are chosen from a list of metals particularly useful in nuclear reactors.

Table 4·3 Fuels and furnaces for melting metals

Fuel	Metals melted	Furnace
Coal.............	Cast iron	Air furnace, cupola
Oil and gas.......	Steel, aluminum, zinc, brass	Crucible furnace, open hearth
Electricity		
arc............	Steel, molybdenum, titanium, zirconium	Electric-arc furnace, inert atmosphere, arc melting
electron........	Refractory and reactive metals	Electron-beam furnace
induction	All metals	Induction furnace
resistance	All metals	Resistance furnace

4·33 Refractories for casting

The *crucible* or container for the molten metal, must be *refractory* and must not react chemically with the hot metal. A *refractory* container is of a heat-resistant material and is generally nonmetallic. Because of the interest in very reactive metals such as uranium, zirconium, titanium, and beryllium, new refractory crucibles have had to be developed. Among the recent developments have been the wider usage of beryllia (BeO), zirconia (ZrO_2), magnesia (MgO), and graphite (C). Table 4·4 lists several properties of common refractory materials, with some pertinent remarks.

When the metal is at the desired pouring temperature, it is poured into a refractory or a water-cooled metal mold. The refractories discussed in the previous paragraph are also used as mold materials. However, the most common mold materials used for cast iron, steel, aluminum, magnesium, etc., consist of sand plus a binder. The mold interior is designed to allow the molten metal to enter the mold (through gates) and solidify without the formation of shrinkage cavities in the useful section of the casting. Students interested in this phase of metal fabrication are advised to do further reading in books on foundry practice.

Table 4·4 Properties of refractories

Material	Formula	Melting point, °C	Resistance to spalling	Remarks
Alumina (fused)......	Al_2O_3	2050	Fair	Used for furnace cores, thermocouple protection tubes, and high temperature insulation
Beryllia	BeO	2530	Good	Used for melting Be, Th, and U
Graphite	C	3500	Excellent	Used for melting U and its alloys
Magnesia	MgO	2200–2800	Poor	
Zirconia	ZrO_2	2700	Fair	

4·34 Powder metallurgy

The powder method provides a means of fabricating highly refractory metals and metal compounds that cannot be cast by the usual melting process. In this process, the powder particles are welded or sintered together, forming a continuous metal phase. Tungsten, molybdenum, and tantalum have melting points beyond the useful limit of any refractories now known; consequently, they must be obtained in the desired form at temperatures below their melting points. The powder method is also used in special situations; specific examples are the production of high-purity beryllium from flake or powder, formation of oxide fuel elements, cemented carbide tools, and production of porous metal bearings. Powder metallurgy is relatively new and has only recently attained a position of commercial importance. However, its possibilities are now being realized, and it is taking its place as an important metallurgical process.

The requisite steps in making a part by powder metallurgy include *powder making*, *classification*, *mixing*, *pressing*, and *sintering*. Metal powders are made by electrolysis, reduction of oxides, atomization, and by mechanical disintegration of molten or solid metal. The powder, usually finer than 0.01 in. in particle size, is *classified* (separated according to particle size) by screening, air separation, and *sedimentation* (settling rate in a fluid). Mixtures of powders differing in size or composition are commonly made in drums with dry lubricants and then pressed in steel or carbide-lined dies at 5 to 25 tons/in.² pressure to form briquettes of self-sustaining shape. Heat-treatment of these briquettes in inert or *reducing* gases (nonoxidizing) markedly strengthens the briquette and in many cases also causes densification. Many products so made are coined to final dimensions or subsequently hot-worked to shape by standard procedures; e.g., bearings are simply pressed mixtures of copper, lead, and tin powders, sintered and coined to final dimensions, whereas a pressed and sintered tungsten rod may be hot-forged, swaged, and drawn into wire.

Sintering is a process of welding adjacent powder particles which depends upon atomic bonding. The actual welding in the process is due to atomic surface forces, which are not dependent on temperature. High temperature acts merely to facilitate the contact obtainable and to increase the mobility of the atoms diffusing across the contact interface.

4·35 Metal-working operations

Subsequent to casting, further shaping operations are frequently desirable, either to produce a new shape or to improve the properties of the metal. For the latter purpose, the metal piece should be reduced to at least two-fifths to one-fifth of its original size. Most metal working carried out in the above manner is performed above the recrystallization temperature. During this high-temperature working, both the phenomena of hardening due to deformation and of softening due to annealing occur simultaneously (see Chap. 3). However, if deformation develops more rapidly than softening can take place at that temperature, the metal will harden. Metals continue to work harden up to a temperature very near their melting points, but the rate of work hardening constantly decreases with increasing temperature. For a given rate of deformation, there exists a definite temperature where hardening is just balanced by simultaneous softening. Above this temperature, deformation is termed *hot working*; below it, *cold working*. Thus, from the standpoint of its effect on the properties and structure of the metal, the temperature at which working is done is not significant unless it is accompanied by some knowledge of the rate at which deformation proceeds. Steel forged at 500°F is said to be *finished cold*, while lead worked at room temperature is being hot-worked.

4·36 Effect of working on metal properties

The properties of a worked metal are controlled by temperature and the rate of deformation. In most practical cases, the desired properties are obtained by very careful control of the *finishing temperature*. The finishing temperature is the temperature at which deformation is completed. With high finishing temperatures (far above the recrystallization temperature), the metal parts will be annealed and may have a large grain size. If, on the contrary, deformation is continued until the temperature of the object is below the recrystallization temperature (low finishing temperature), the metal will be hard and will show directional properties.

In addition to producing desired shapes most economically, hot working improves the properties of the as-cast metal. Not only are there a refinement of grain size and a partial elimination of cast structure, but deformation and annealing also crack up the shells of impurities at grain boundaries and permit the elimination of crystal segregation or cored structure by diffusion. A simple example of the improvement possible by hot working is shown in Fig. 4·44. Two hooks are sketched to show the direction of planes of weakness. Splitting takes place most readily parallel on these planes rather than at right angles to them. Thus, the hot-worked hook is considerably stronger and more impact-resistant than the hook cut from a cast section.

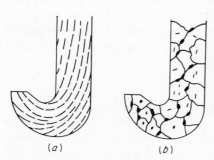

(a)   (b)

Fig. 4·44 Two hooks, one formed by hot bending, the other cut from a cast billet. (a) Impurities are strung out by hot working. (b) Impurities at grain boundaries.

4·37 Working operations— compression

Analysis of the numerous processes used in industry for working metals reveals that most of them fit moderately well into one of four general classes, viz., those based on compression, tension, shearing, or bending. Each of these metal-working classifications will be discussed briefly, and in addition important variations particularly applicable to the production of reactor fuel elements will be described.

Fundamentally, all compression operations are variants of simple reduction in thickness of a block of metal when struck with a hammer or squeezed between the platens of a press. Included in this classification are the hammer operations of forging, swaging, certain forms of extrusion, and rolling. These operations, pictured schematically in Fig. 4·45a, account, on a tonnage basis, for over 85 per cent of metal-working processes.

Fabrication of a large number of reactor fuel elements requires the use of at least one of the operations given in Fig. 4·45a. Two such fuel elements will be discussed briefly. The first to be described is the "canned-slug" type of element shown in Fig. 4·46a. It consists of a uranium core surrounded by a cladding material, such as aluminum, stainless steel, or zirconium. This type of element can be manufactured with or without a *metallurgical bond* between the core and cladding. In this example, a metallurgical bond can be defined as a zone of interdiffusion between two metals in contact. Without this bond, the fuel-element process consists in separately shaping the core (by casting and machining) and the cladding (by extruding a tube), combining them, and sealing the ends with welded end caps. The metallurgically bonded elements are fabricated by a process called "coextrusion." The core

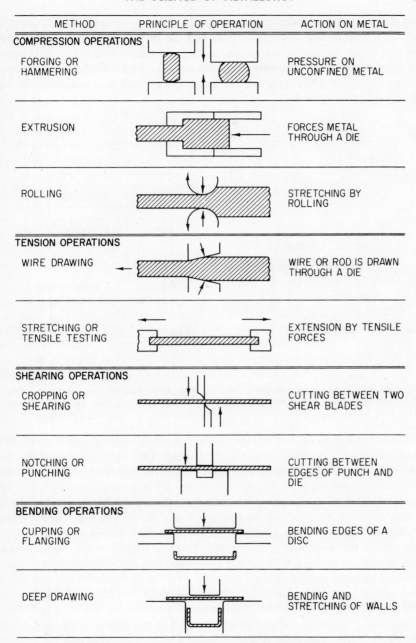

METHOD	PRINCIPLE OF OPERATION	ACTION ON METAL
COMPRESSION OPERATIONS		
FORGING OR HAMMERING		PRESSURE ON UNCONFINED METAL
EXTRUSION		FORCES METAL THROUGH A DIE
ROLLING		STRETCHING BY ROLLING
TENSION OPERATIONS		
WIRE DRAWING		WIRE OR ROD IS DRAWN THROUGH A DIE
STRETCHING OR TENSILE TESTING		EXTENSION BY TENSILE FORCES
SHEARING OPERATIONS		
CROPPING OR SHEARING		CUTTING BETWEEN TWO SHEAR BLADES
NOTCHING OR PUNCHING		CUTTING BETWEEN EDGES OF PUNCH AND DIE
BENDING OPERATIONS		
CUPPING OR FLANGING		BENDING EDGES OF A DISC
DEEP DRAWING		BENDING AND STRETCHING OF WALLS

Fig. 4·45   Pictorial survey of metal-working operations.

and clad sections are carefully cleaned and sealed in vacuum inside a copper can, as shown in Fig. 4·46b. The nose and tail of the core-cladding interface are preshaped by machining so that a flat interface is formed by extrusion. By careful control of the extrusion temperature and the reduction of area during extrusion, a fuel element will be produced with the core and cladding metallurgically bonded.

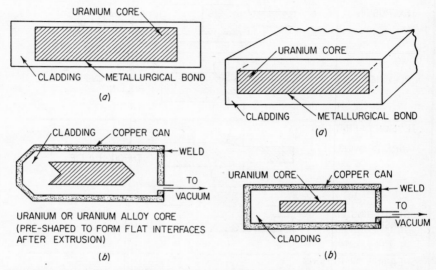

Fig. 4·46   Extruded fuel element. (a) "Canned-slug" fuel element; (b) "co-extrusion billet."

Fig. 4·47   Rolled fuel element. (a) "Picture-frame," or "sandwich," fuel element; (b) rolling assembly.

The second type of fuel element is fabricated by rolling and is called the "picture-frame," or "sandwich," fuel element. This element is shown schematically in cross section in Fig. 4·47a. The picture-frame element is assembled and enclosed in a copper can and evacuated as shown in Fig. 4·47b. The technique of surrounding the element with copper or iron before fabrication is common practice, to prevent oxidation of the core or cladding during the heating which generally precedes the fabrication operation. After evacuation the assembly is hot-rolled. Control of the rolling temperature and reduction of area will permit the formation of a metallurgical bond between the core and cladding.

4·38   Other working operations

Tension operations are not so widely used as the compression operations described above. Drawing, whereby metal is pulled through dies or between idling rolls, is considered in this classification, although drawing may also be viewed as compression of metal by the dies or rolls. Operations more strictly tensile in nature are stretching for purposes of straightening, stretching as part of certain bending operations, and dieless wire drawing, a process still at the experimental level. The tensile test, used for determining mechanical properties of metals, is the prototype of tension operations. The pictorial survey of metal-working operations given in Fig. 4·45 includes tension operations.

Shearing operations of various kinds are used at several stages in the manufacture of most articles, from the time the ingot is cropped in the blooming, or slabbing, mill until the finished article receives its final polish. The basic operation is *cropping*, or shearing a bar between two shear blades. But the same fundamental mechanism is found in all metal-cutting operations: in sawing, filing, grinding, in cutting on a lathe or mill, in drilling, boring, etc. Shearing operations performed on presses are blanking, punching and notching, and slitting and trimming; these are generally applied to finished products, such as strip, sheet, and bar stock.

Finally, the fundamental process of bending is a part of many forming operations. There are two variants, which can be classified as bending a curved flange and bending a straight flange. For sheet metal, *plain bending* (bending a straight flange) is done on various machines: presses, brakes, rolls, draw benches, and other more specialized machines. Operations resembling bending a curved flange are performed on presses or hammers (deep-drawing) or on lathes (spinning).

The production of an article may require several operations classified under different headings; also, there are usually several ways of producing the same article. For example, tubes may be made from round stock by several compression operations (roll piercing, extrusion, or impact extrusion) or by shearing (drilling or boring), or they may be made from strip by various bending operations (deep drawing, spinning, roll bending, or roll forming). The owner of a rolling mill and the owner of a plant equipped only with presses can produce many identical products by entirely different methods.

4·39 Welding

Welding, brazing, and soldering are processes in which parts are joined by means of the forces of interatomic attraction across the joint. Riveting, bolting, and other mechanical methods of joining parts depend not upon atomic forces to act across the joint, but simply upon the rigidity of the parts and of the connectors. Thus, a riveted joint may connect two steel surfaces, each surface being coated with scale or rust, so that there is no intimate contact across the joint and no atomic attraction between the contact surfaces. Welds, on the contrary, require clean, intimate contact at the joint, so that atoms in opposite faces come close enough to attract each other and bind or weld the parts. Figure 4·48 gives a chart of welding

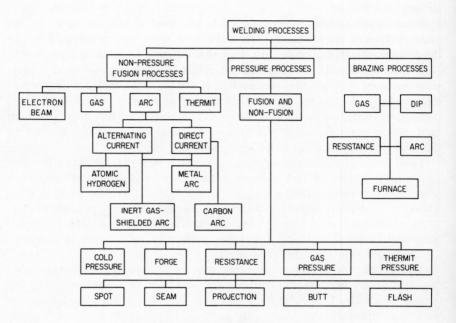

Fig. 4·48   Chart of welding processes.

processes which has three subdivisions: fusion welding, pressure welding, and brazing. We shall examine briefly the principles of each of these three welding techniques.

4·40   Fusion welding

A fusion weld is made by applying heat to abutting metal members until the mating parts actually fuse, or melt together. Such heat is usually derived from a point source (for example, an electric or gas torch) that is steadily advancing along the joint to be made. The heat absorbed from the hot source spreads into the adjacent metal, causing a rise in temperature. When the metal is molten just beneath the heat source, the heat source is advanced to fuse the remainder of the joint. As the heat source moves on, the molten weld metal begins to lose heat to the parent metal and to the air and gradually solidifies. This sequence of melting and solidification is shown in cross section in Fig. 4·49.

Surrounding the melted region there is a layer of coarse grains. These coarser grains are formed when the fine-grained parent metal is heated near its melting point, at which temperature grain growth takes place. If we examine the grain structure farther from the melted region, we see the grains that were unaffected by the welding process. During solidification, the molten pool freezes as long, thin, columnar grains which grow outward from the base metal.

Although fusion welding is a typical melting and casting procedure and equilibrium phase diagrams are useful in interpreting the behavior of weld

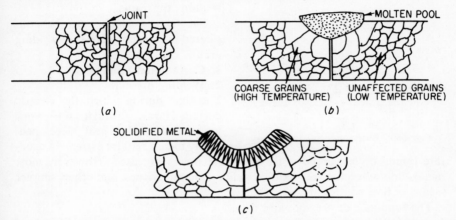

Fig. 4·49 Sequence of melting and solidification in fusion melting. (a) Grain size before welding; (b) grain growth in vicinity of the joint; (c) final welded joint.

and base metal, the conditions are strictly of a "nonequilibrium" nature due to the short time cycle involved. Other complicating and at times harmful developments during fusion welding are:

I. Gas absorption
2. Slag inclusions
3. Segregation and shrinkage during solidification
4. Hot cracks

Most of these difficulties may be controlled by definite, well-regulated practice; for example, methods of reducing gas absorption in weld metal are:

I. Provision of an inert protective atmosphere above the weld
2. The use of fluxes that form protective slags on the molten metal surface
3. Avoiding overheating

4·41 Pressure welding

Pressure welding is the joining of metal parts by forging them together under pressure; to aid fusion at the interface, the parts are heated to the plastic condition. Of all pressure welding techniques, resistance welding is most widely used. The metal is heated by passing an electric current perpendicular to the weld interface. The high contact resistance at the interface concentrates the heating in this area. The pressure applied during the heating cycle causes the parts to join in the plastic state by interdiffusion and atomic attraction. In many cases, actual liquid fusion takes place, at least in a small portion of the weld area where the temperature reaches a high local level. Figure 4·50 shows the general characteristics of a resistance spot-welding apparatus.

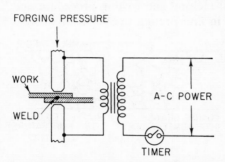

Fig. 4·50   Resistance spot welding.

The factors which must be considered in resistance pressure welding are:

1. Current and timing

2. Welding pressure

The time during which the electric current flows through the work governs the total heat developed. Many metals require extremely accurate timing cycles, and special electronic timers are used. However, most metals do not require such critical welding characteristics, and other, simpler means of time control are satisfactory.

The functions of pressure are:

1. It breaks down oxide film on the surface layers of materials to be welded and forces the metals into intimate contact at the interface.

2. By pressing the pieces together, the passage of current is limited to this area.

3. Pressure reduces the formation of porosity and cracks in the weld area.

4. Deformation occurs in the crystals of metals being welded and induces recrystallization and grain growth across the interface.

The cycle of operations is usually timed so that the metals to be welded are under pressure prior to the application of current. The pressure is maintained for a short time after the timer has interrupted the current flow.

4·42   Brazing

Brazing covers a group of welding processes in which metallic parts are joined by a liquid, nonferrous metal or alloy whose melting point is above 535°C (1000°F) but below that of the metals being joined. The various methods of brazing are named for the type of filler metal used or from the means of applying heat: copper or silver brazing, furnace or induction brazing, etc. *Soldering* is a low-temperature form of brazing and may be described as the joining of metals with low-temperature filler material, usually alloys of lead and tin.

Table 4·5 Typical brazing and soldering alloys

A. Brazing Alloys

Metal or alloy	Composition				Temperature, °F		Uses and remarks
	Cu	Zn	Ag	Others	Solidus	Liquidus	
Copper	100	...	..	....	....	1982	Hydrogen atmosphere furnace brazing
ASTM No. 2......	45	35	20	....	1430	1500	For copper, copper alloys, nickel, silver, steel, and iron
ASTM No. 6......	20	15	65	....	1280	1350	Excellent for Monel
Easy Flo..........	15.5	16.5	50	18Cd	1160	1175	Ferrous and nonferrous

B. Soldering Alloys

	Pb	Sn	Ag			
Silver-lead	97.5	..	2.5	580	580	High strength at elevated temperatures
50–50	50	50	...	360	420	The preferred general-purpose solder
Soft	37	63	...	360	360	Eutectic composition, high fluidity

The characteristics of a truly brazed joint are:

1. Complete wetting of parts by the molten filler metal
2. Complete filling of space in the joint
3. A limited amount of alloying between the filler and base metals

The principal forces acting in brazing are those of *surface tension* and *capillary attraction.* Just as any liquid will rise in capillary tubes when such tubes are cleaned and set into a reservoir, so will the liquid filler metal rise against gravity in spaces between metal parts to be joined. The success of this operation will depend upon:

1. The ability of the brazing alloy to wet the surfaces to be joined
2. The viscosity of the alloy in the liquid state
3. The weight of the column of liquid required to fill the space

A summary of some properties and uses of common brazing and soldering metals and alloys is given in Table 4·5.

PROBLEMS

Note the correct statements; there may be none, one, or more than one.

I. The field of process metallurgy includes such processes as (a) ore dressing; (b) refining; (c) rolling; (d) melting.

2. In the Hall process for aluminum, the significant features of the operation are (a) electrolysis; (b) fused mixture of Al_2O_3 and Na_3AlF_6; (c) tungsten electrodes; (d) liberation of oxygen from the cathode.

3. Pyrometallurgy methods include (a) roasting of sulfides; (b) blast-furnace operations; (c) leaching; (d) open-hearth operations.

4. Steel is an alloy of iron plus small but carefully controlled amounts of (a) carbon; (b) nitrogen; (c) hydrogen; (d) chlorine.

5. Metal replacement and halide decomposition are processing techniques for important reactor materials such as (a) plutonium; (b) zirconium; (c) palladium; (d) uranium.

6. Typical examples of alloys include (a) brass; (b) aluminum-lead; (c) steel; (d) cast iron.

7. (a) Aluminum-lead alloys are completely miscible in both the liquid and the solid state; (b) small additions of aluminum to lead lower its melting point considerably; (c) the cooling curves of aluminum-lead alloys show two temperature holds; (d) at room temperature, the microstructure of aluminum-lead alloys would be single-phase.

8. All binary metal phase diagrams contain (a) a solidus; (b) a eutectic temperature; (c) a liquidus; (d) a peritectic temperature.

9. The eutectic reaction is characterized by (a) a eutectic temperature; (b) a eutectic phase; (c) a eutectic mixture; (d) a eutectic composition.

10. The peritectic reaction (a) is very rarely encountered in alloy systems; (b) is characterized by a peritectic temperature; (c) is always found in systems showing complete solubility in both the liquid and the solid state; (d) is represented by the following reaction on cooling:

Liquid + alpha (solid solution) → beta (solid solution)

II. An ordered solid solution is characterized by (a) government specifications; (b) ASM specifications; (c) particular lattice positions for each type of atom; (d) the ordered phase copper-gold.

12. The hardness of a series of binary alloys (a) is a linear relationship with composition; (b) is independent of composition; (c) is a complicated relationship depending in the interactions of the different atomic species; (d) is predictable by modern metallurgical theories.

13. Precipitation-hardening reactions are characterized by (a) decreasing solid solubility with decreasing tempreature; (b) increasing solid solubility with decreasing temperature; (c) increasing solid solubility with increasing temperature; (d) none of these.

14. Heat-treatment is used to carry out the following changes in properties of metals: (*a*) to harden a precipitation-hardening alloy; (*b*) to anneal an extruded rod; (*c*) to change the density of an annealed metal; (*d*) to recrystallize a cold-worked metal.

15. A dislocation (*a*) is a crack in a metal crystal; (*b*) is the boundary between slipped and unslipped regions of a slip plane; (*c*) is often represented by an extra row of atoms which ends at the slip plane; (*d*) is necessary to explain why the observed critical shear stress in metals is not as high as the theoretical value.

16. A lattice vacancy is (*a*) a visible void in a metal; (*d*) the hole which exists when a lattice site is left vacant; (*c*) an imperfection in metal crystals; (*d*) the space left between perfect, close-packed rows of atoms.

17. The field of fabrication metallurgy includes such processes as (*a*) aging; (*b*) solution treatment; (*c*) swaging; (*d*) deep drawing.

18. Powder metallurgy is (*a*) a coating process using a fine spray of powdered metals; (*b*) an ancient method of fabricating metal parts by melting fine metal powders; (*c*) a means of fabricating highly refractory metals and metal compounds that cannot be cast by the usual melting process; (*d*) a process related to the production of aluminum and titanium oxide paints.

19. Examples of compression-working operations are (*a*) rolling; (*b*) shearing; (*c*) drawing; (*d*) forging.

20. Brazing and soldering are processes for joining metals by (*a*) a non-ferrous filler metal; (*b*) fusion of the metal parts; (*c*) arc welding; (*d*) none of these methods.

21. Construct the Pb-Sn phase diagram from the following information: The melting point of Pb is 327°C and that of Sn 232°C. A eutectic is found at 183°C and 62 per cent Sn. At this temperature, Pb dissolves 3 per cent Sn. Assume that the solubility of Pb in Sn and of Sn in Pb at room temperature is 1 per cent. (*a*) What per cent of an alloy of 25 per cent Sn and 75 per cent Pb contains eutectic at 183°C and room temperature? (*b*) What is the composition of the phases present in the above alloy at 183°C and at room temperature? (All values are in weight per cent.) (*c*) Draw the microstructures of alloys containing 16 per cent Sn, 25 per cent Sn, and 62 per cent Sn.

22. From the U-Cr phase diagram given in Fig. 4·19, answer the following questions: (*a*) How many eutectic reactions are there? (*b*) How many eutectoid reactions? (*c*) What are the phases present in an alloy containing 2 per cent Cr at 800°C, 10 per cent Cr at 645°C, and 50 per cent Cr at 600°C?

23. From the Cu-Ni phase diagram given in Fig. 4·14: (*a*) How would you classify this type of binary system? (*b*) In an alloy containing 50 weight per cent Cu, what is the approximate composition of the phase or phases present at 1400, 1300, and 1200°C? Also, what is the relative amount of each of the phases present at these temperatures? (*c*) Draw the microstructure of the 50 per cent Cu alloy at room temperature.

5

PROBLEMS IN MATERIALS FOR NUCLEAR POWER

PART I. GENERAL PROBLEMS

5·1 Introduction

Within the period 1956 to 1960, the USAEC invested $200 million in a program to develop economical nuclear power plants. The first round of reactor building, which included the MTR, STR I, and STR II (pressurized water reactors), the BWR (boiling water reactor), the HRE I (homogeneous reactor), and the EBR I (breeder reactor), has been essentially completed. The second round of reactor construction, which is just under way, will be concentrated on five designs, each of which is an improvement or variation of one of the first-round reactors.

It might be asked why the USAEC is studying five separate approaches, and not just one, or even twenty. The answer is that no one line of reactor development has yet given convincing evidence of superior promise. In addition, the various reactor types are not really competitive designs, because the objectives of each are not entirely comparable. On the other hand, not even the United States is rich enough to investigate every promising possibility. Within nuclear power groups, it is a common joke that there are as many ideas for reactors as there are reactor designers. The five types of reactors comprising the second round have been explored extensively for several years and have been selected as the most promising.

We shall consider some of the more general materials problems of the five types of reactors to be investigated. As an aid to this discussion, most of the pertinent data for each reactor are listed in Table 5·1. The first four reactors described utilize solid fuel elements, while the last has a fluid-fuel system. The fluid-fuel system has several important advantages over solid fuel elements, as we shall see, but these advantages are balanced by a new set of problems. Each of these two broad types of fuel systems gives rise to rather different sets of problems, which will be taken up successively.

Table 5·1   Data on nuclear reactors in the power development program

Pressurized water reactor (Westinghouse):

Demonstration ......PWR, Westinghouse

Fuel system ....Enriched uranium-zirconium alloys and uranium oxide, each contained in zirconium jackets

Moderator ....H_2O at 2,000 psi

Heat transfer ....Circulating water at 2,000 psi, 525°F, to boiling water at 486°F

Container ....Stainless-steel-lined steel pressure vessels, stainless-steel piping

Pumping............Canned rotor and mechanical pumps

Problems ....Cheaper fuel fabrication, better neutron economy, higher thermal efficiency through higher water pressure

Boiling water reactor (Argonne):

Demonstration ......BEWR, Argonne

Fuel system ....Slightly enriched uranium alloys or compounds contained in zirconium (or possibly aluminum)

Moderator ....Water [graphite in a General Electric (GE) variation]

Heat transfer ....Boiling water converted directly to steam power

Container ....Similar to pressurized water reactor

Pumping............Condensate return only by canned rotor

Problems ....Stability of control during boiling and similar materials problems as in pressurized water reactor

Sodium-graphite reactor (Atomics International):

Demonstration ......SRE, Atomics International

Fuel system ....1. Slightly enriched uranium
                     2. Thorium-uranium alloy, contained in Na- or NaK-filled stainless-steel tubes
                     3. Uranium carbide

Moderator ....Graphite

Heat transfer ....Circulating sodium to circulating NaK to water and steam at 800 to 900°F

Container ....Stainless steel (e.g., type 304)

Pumping............Mechanical and freeze seal

Problems ....Cheaper plumbing (piping and pumping) for sodium and a radiation-damage-resistant fuel

Fast breeder reactor (Argonne):

Demonstration ......EBR-I, EBR-II, Argonne

Fuel system ....1. Highly enriched uranium or
                     2. Uranium-plutonium alloys, both contained in NaK- or Na-filled steel tubes. The fuel to be surrounded by a blanket of depleted uranium

Moderator ....None

Heat transfer ....Circulating NaK or Na to circulating NaK to water and steam

Container ....Stainless steel

Pumping............Electromagnetic

Problems ....Heat removal from fuel and radiation-damage-resistant fuel

Table 5·1   Data on nuclear reactors in the power development program (cont'd)

Homogeneous aqueous reactor (Oak Ridge):
    Demonstration ......HRE-I, HRE-II, HTRE, Oak Ridge
    Fuel system ....1. UO_2SO_4 dissolved in D_2O
                         2. For HTRE the core is blanketed with a thorium salt
                            in D_2O
    Moderator ....D_2O
    Heat transfer ....Circulating fuel to boiling water at approximately 450°F
    Container ....Stabilized stainless steel
    Pumping ....Canned rotor
    Problems ....Corrosion and handling of highly radioactive fuels

5·2   Solid-fuel reactors

Many of the problems in reactor materials may be outlined by inspecting closely some typical, representative solid-fuel reactors and noting the materials used and the conditions to which they are subjected. Instead of examining any one of the reactor designs in detail, it will be desirable to examine in a general way problems which will be the same for all solid-fuel reactors. We shall start this discussion by focusing our attention on the temperature distribution in the heart of the reactor.

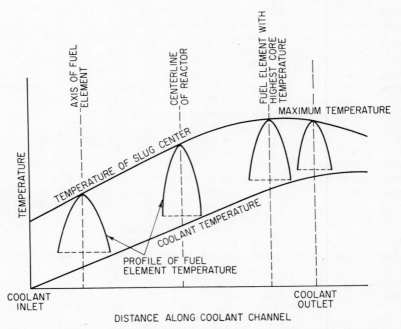

Fig. 5·1   Temperature distribution along a coolant channel and across fuel elements. The temperature profile of several elements is given.

Figure 5·1 shows the general case of temperature distribution when the coolant enters at one end, is heated by the fuel elements, and leaves the reactor at a higher temperature. The neutron flux and the fission rate in the fuel elements, in general, increase from the edge to the center of the reactor. There will therefore be a fuel element just downstream from the center of the reactor in which the central portion of the element is at the highest temperature and the surface temperature is nearly as high as the outlet coolant temperature. Table 5·2 gives some of the details of the temperature values in several reactors.

Table 5·2   Representative temperature values in solid-fuel power reactors

Reactor	Coolant temperature		Maximum core temperature $T_{central}$, °F	Power flux, Btu/ft²-hr
	T_{in}, °F	T_{out}, °F		
EBWR......	110	488	610	130,000 (maximum)
PWR	508	542	1070	118,000 (average)
SRE........	500	960	1200	340,000 (maximum)
EBR-I	822	1000	1390	1,300,000 (maximum)

The detailed anatomy of a solid fuel element will serve as a model for examination of the major reactor materials problems. The schematic cutaway section of the fuel element and coolant is shown in Fig. 5·2. The temperature gradient through the section is also indicated. The important phenomena, which in turn establish the conditions prevailing in the element, are as follows:

1. The fission event.
2. The kinetic energy of the fission recoil is dissipated to the crystal lattice, producing defects usually called radiation damage.
3. The fission product atoms take up positions close to the end of their recoil range in the material or diffuse to more stable configurations, depending on the temperature. Since the atomic volume of the fission product atoms is greater than that of the uranium atom, a corresponding increase in the fuel-element volume must be expected. If the temperature is high enough to permit diffusion of the inert-gas fission products, nucleation and growth of gas pockets must be expected. Both these phenomena will contribute to compressive forces within the fuel element and will cause tensile forces within the cladding material.
4. All but a small fraction of the fission energy appears as heat in the bulk material and flows through the cladding to the coolant.
5. The steady-state temperature and temperature gradients set up determine:
   a. Stresses and strains.
   b. Reactions between cladding and coolant.

   c. Interdiffusion between fuel and cladding.

   d. Diffusion of fission products.

With this schematic model of a fuel element in mind, we shall examine the types of problems encountered by reactor designers. Starting from the center of the fuel element and working toward the coolant, the important problems are expected in the following locations:

I. Inside the fuel, radiation damage and accumulation of fission products

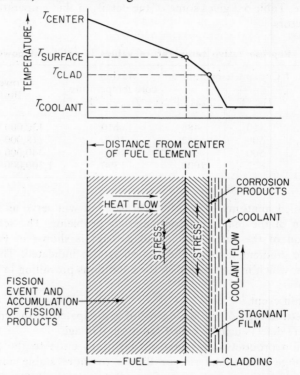

Fig. 5·2   Schematic cutaway section of fuel element and coolant.

2. Between the fuel and cladding

3. Between the cladding and the coolant

4. Between the fuel and coolant in case there is an imperfection in the cladding

5. In the coolant itself

In the next few sections, we shall examine many of the details of these problems. The requirements for an understanding of the problems, current experiments, and present designs are a knowledge of both metallurgy and solid-state physics. The student is advised to consult sections of the last two chapters when language or concept difficulties first arise. Many problems, however, will not yield to this technique but will require appreciable outside reading.

5·3 Fluid-fuel reactors

Many of the problems associated with solid fuels and claddings are replaced by a quite different set when fluid fuels are used. Radiation damage to the fuel is absent. There is no problem of transfer of heat from the fuel to the coolant, and the high internal temperature of solid fuel elements has no parallel in the case of fluid fuels. Moreover, fission products may be removed at will, continuously or semicontinuously, so that the problem of their buildup may be completely avoided. Problems of corrosion of the container, piping, and heat-exchanger materials are increased compared with the solid-fuel system, and in particular the problem of contamination of the entire system with radioactive fission products.

To compare solid- and fluid-fuel systems, in the solid-fuel system the parts which wear out, namely, the fuel elements, are removed and replaced. The remaining parts of the system can be designed and fabricated for a reasonably long life. In the case of fluid fuels, the whole system has a life governed by the amount of corrosion; moreover, the reactor is part of the fuel-handling and -recovery system. Areas of severe corrosion can be made more rugged than in solid-fuel reactors, but they cannot be made thicker than the amount permitted by efficient heat-transfer design. A comparison of solid- and fluid-fuel reactors should list some of the economics involved, such as fuel-handling costs of solid-fuel reactors, replacement of fluid-fuel systems, maintenance, and fuel inventory. A discussion of these factors is outside the scope of this chapter, but this information is available from the Reactor Development Program of the USAEC.

In the case of the aqueous homogeneous reactor, described in Table 5·1, materials problems arise primarily at the interface between the fluid and the container material. The fluid fuel is a solution of uranylsulfate in D_2O at rather low pH (pH is a measure of acidity or basicity; low pH means an acid solution). Other than to say that both stainless steel and zirconium are being considered for this application, the problem is much too complicated to be summarized here. It does, however, illustrate the type of research which must be carried out in the development of reactors, namely, research on aqueous corrosion at elevated temperatures.

One of the early reactor concepts, development of which was deferred because of the lack of engineering and scientific information, is the liquid-metal-fuel reactor. This reactor and the accompanying materials problems are being studied at the Brookhaven National Laboratory. The data design are listed in Table 5·3.

Table 5·3   A liquid-metal-fuel reactor

Demonstration ......None
Fuel system ....Enriched uranium dissolved in bismuth
Moderator ....Graphite
Heat transfer ....Circulating fuel at about 500°C to water and steam
Container ....Ferritic high-temperature steel, e.g., Croloy
Pumping ....Electromagnetic or mechanical
Problems ....Corrosion and handling of highly radioactive liquid metals

The materials problems, more metallurgical than those in the case of the aqueous homogeneous reactor, may be summarized as follows:

1. Determination of solubility of fuel materials in liquid metals, particularly bismuth
2. Corrosion of materials of construction by liquid metals, and means of inhibiting corrosion
3. Means of continuous extraction of fission products from molten bismuth solutions
4. Methods of preparing more concentrated liquid-metal fluids containing thorium, to serve as the fertile material in such reactor systems

The first two of these problems have almost been solved by the staff at Brookhaven. The solubility of uranium in bismuth has been found to be sufficiently high so that a solution 800 ppm may be maintained at desirable operating temperatures around 500°C. The problem of corrosion of stainless steels by liquid bismuth was studied extensively. Using radioactive tracer techniques, among other specialized methods, the Brookhaven staff has found that small additions of zirconium to the bismuth inhibit the corrosion of iron and that austenitic stainless steels may not be used as containers for high-temperature bismuth solutions because of the formation of stable nickel-bismuth intermetallic compounds and the gradual diffusion of nickel from the steel.

PART 2.   RADIATION DAMAGE

5·4   Preview of Parts 2, 3, and 4

In order to better understand some general materials problems, we shall examine in detail three principal sources of problems: radiation damage, corrosion, and temperature. We shall examine the problem of radiation damage first, for several reasons: (1) Solid fuel elements, the core of the reactor, are very sensitive to radiation damage, and this sensitivity controls the cost of power from nuclear reactors, because it limits the allowable burnup before reprocessing becomes necessary. (2) All materials within the reactor are being continuously irradiated; thus, radiation damage can be a possible problem in any reactor material. (3) Radiation damage can affect the corrosion rate and add to some of the thermal problems within a reactor.

(4) The problem is a brand-new one for most engineers and requires a rather detailed description. After the discussion of radiation damage, we shall examine in somewhat less detail some corrosion and thermal problems.

5.5 Simplified theory of radiation damage

The effects of high-energy radiation on matter, on metals in particular, have been under intensive study since the construction of the first nuclear reactors. E. P. Wigner first called attention to the fact that the energetic neutrons, born in the fission process, would have the ability to displace atoms from equilibrium positions in the crystal lattice of solids that might be far removed from the fuel. As a consequence the neutrons would have deleterious effects on many properties of engineering interest in reactor construction. This observation, plus the obvious realization that considerable damage to the fuel would result from fission fragments during fissioning, prompted an immediate program of theoretical and experimental study of the magnitude of the effects to be expected.

Early experimental studies, followed up by theoretical analyses by F. Seitz, G. J. Dienes, and H. Brooks, confirmed Wigner's prophecy and led to suitable action by the reactor designers to allow for the damage expected to result from reactor operation. Furthermore, it soon became obvious that radiation damage was to be expected in all solid materials and that the susceptibility of various materials to neutron and fission fragment effects would be a function of that particular solid. It was also apparent that the magnitude of the effect was a function of the neutron flux, the temperature, the environment during irradiation, and other variables such as stress.

The above observations led the USAEC to establish research programs designed to study the basic mechanisms of radiation damage; to make comprehensive studies of the effect of neutron irradiation on the physical properties of various solids; and to survey, in so far as possible, radiation damage in a wide variety of materials of interest under conditions appropriate to possible use. This program has been active for many years. While progress has been considerable, it is not now possible to give the reactor designer all the information he needs in a quantitative form. However, it is recognized that it will be possible eventually to derive engineering formulas that will enable a designer to calculate probable effects as a function of time, flux, and temperature.

Radiation damage in a solid may arise in three ways:

1. Through interaction of fission fragments with atoms of the parent-fuel lattice. This type of damage is usually confined to the fuel-bearing material, because the range of the fission fragments is only of the order of microns even in the most transparent solids (in graphite, the range is less than 20 μ).

2. Through elastic collision of a neutron with atoms of a lattice. This type of damage may occur at relatively great distances from the source of the neutrons, since the probability of capture of a fast neutron is very low at high neutron energies. This means that parts of a reactor other than the fuel will be affected, and hence the moderator, shield, or structural components of the reactor can be damaged.

3. Through ionization effects caused by fission fragments or primary recoil atoms traversing the material or by beta and gamma rays. Ionization effects will be discussed very briefly, since they are of consequence only in those solids which are of auxiliary value to a reactor such as plastics.

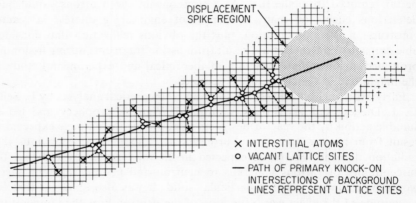

DISPLACEMENT SPIKE REGION

× INTERSTITIAL ATOMS
○ VACANT LATTICE SITES
— PATH OF PRIMARY KNOCK-ON
INTERSECTIONS OF BACKGROUND
LINES REPRESENT LATTICE SITES

Fig. 5·3 Path of primary recoil atom and lattice damage produced.

Except for fuel elements which exhibit damage caused by fission fragments, the principal cause of radiation-damage effects in metals is displaced atoms. On an atomic scale, the production of displaced atoms may be considered by first examining the average energy transmitted in a collision between an energetic neutron and a lattice ion. The average transmitted energy is given by

$$\overline{\Delta E} = \frac{2E_0 mM}{(M + m)^2} \tag{5·1}$$

where E_0 = energy of neutron
M = mass of ion
m = mass of neutron

For example, when a 2-mev fission neutron collides with an iron atom, a primary recoil atom of 60 kev is created. The atomic events which are believed to occur as the recoil atom moves through the lattice are given in Fig. 5·3.

When the energy of the primary recoil atom is greater than a certain threshold value, it loses energy by creating low-multiplicity vacancies and interstitials and by ionizing electrons of the metal. When the energy is below the threshold value, the mean-free path between elastic collisions becomes of the order of one atomic spacing and a region of multiple displacements is formed. The multiple-displacement region has been named a "thermal spike" by F. Seitz and a "displacement spike" by J. A. Brinkham. The conceptual differences between a thermal and a displacement spike will be discussed briefly in Sec. 5·7.

The physical consequences of the above interactions are that several types of defects are introduced which affect the properties of a solid. These defects are vacant lattice sites, interstitial atoms, thermal or displacement spikes, ionization effects, and impurity atoms (fission fragments and impurities created by neutron capture). Each of these radiation-induced defects causes property changes in solid materials. Some of these property changes will be described in detail in the following sections.

5·6 Vacancies and interstitials

Experiments indicate that, on an atomic scale, primary recoil atoms will create knocked-on atoms, which leave behind vacant lattice sites and finally come to rest in interstitial positions. Vacancies and interstitials will be introduced in random clusters; within the clusters, the concentration will be higher than calculated on the basis of the total number of displaced atoms. The vacancy-interstitial pairs are believed to affect the hardness, electrical resistivity, and critical shear stress. The presence of vacancies and interstitials may also affect the rate of kinetic processes, which involves micro-diffusion, such as nucleation to precipitate particles in supersaturated solid solutions.

The increase in resistivity of metals due to radiation damage can be explained by the vacancy-interstitial model. Resistivity is generally considered to be made up of two parts, one dependent on temperature and the other dependent upon lattice imperfections such as vacancies, interstitials, impurities, and dislocations. The equation relating the two parts is given by

$$R = R_0 + R_T \qquad (5·2)$$

where R_0 is the residual resistivity caused by lattice imperfections and R_T is the resistivity caused by thermal vibration of the atoms. Experimental work has shown that the residual resistivity R_0 of copper wires irradiated at $-150°C$ increased by a factor of 6 after neutron bombardment.

5·7 "Thermal" or "displacement" spikes

Dissipation of the energy of the fast particle may be viewed as being caused by the introduction of a large amount of heat at a localized point in the material, the region surrounding this point being heated to a high temperature. Calculations by H. Brooks and others have indicated that the possible duration of a high-temperature region of approximately 1000°K (727°C) involving some 5,000 atoms might be 10^{-10} to 10^{-11} sec. The result is a so-called thermal spike, i.e., rapid heating and quenching of a small volume of the material. A fraction of the atoms will be left in displaced positions after the cooling of a spike, and the end result in pure metals is equivalent to a localized production of vacancies and interstitials. In alloys

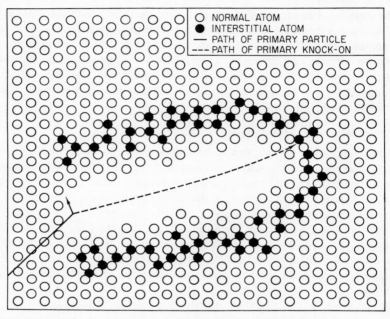

Fig. 5·4   Conceptual representation of a displacement spike.

or compounds, the rapid heating and quenching of the material may promote local atomic exchanges and thereby leave the substances in an altered state, even though most of the atoms have returned to lattice positions. The best-known example is the disordering of ordered alloys by irradiation.

A more detailed model of the lattice damage resulting from the passage of an energetic knock-on (recoil) atom through a solid has been described by J. A. Brinkman. A schematic diagram of the sequence of events following such an energetic collision has been shown in Fig. 5·3. The highly energetic primary recoil atom makes elastic collisions with atoms of the parent lattice, leaving behind a trail of vacancies and interstitials. Between displacement collisions (which produce vacancies and interstitials), energy is lost by inelastic and weak elastic collisions and ionization, which result merely in a tiny wake of heated material, sometimes called a thermal spike. It must be made clear to the student that this thermal spike is different from the one discussed in the preceding paragraph. Unfortunately, Brinkman used the term which identifies Seitz's concepts of radiation damage. Brinkman has estimated that the wake of heated material would cause a local rise in temperature of about 40°C.

Finally, when the energy of the fast-moving recoil atom falls below a transition value which depends on the atomic number, the mean-free path between displacement collisions becomes of the order of the interatomic spacing and each collision results in a displaced atom. Figure 5·4 shows a conceptual representation of how atomic displacement collisions become less than an interatomic distance. The end of the trail is believed to be a region containing of the order of 1,000 to 10,000 atoms in which local melting and turbulent flow have occurred during a very short interval. This last region has been termed a displacement spike.

For purposes of theoretical calculations, Brinkman has further simplified the model as shown in Fig. 5·5. The displacement spike, lasting for much

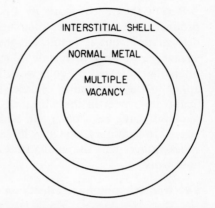

Fig. 5·5 Very simplified model of a displacement spike.

less than a billionth of a second, is considered to be a multiple vacancy surrounded by normal metal, which is in turn surrounded by an interstitial shell. The pressure in the interstitial shell then collapses the normal metal shell and the atoms move to more stable positions. Brinkman visualizes the formation of dislocations as well as vacancies and interstitial atoms during the local recrystallization within the spike region.

5·8 Comparison of radiation damage and cold work

Early in the study of radiation damage, several scientists suggested that there was an analogy between radiation damage and cold work. This analogy was based on the notion that the properties of crystals were a unique function of atomic positions, regardless of how the atoms got there. Experiments had shown that many irradiation effects were similar to those produced by cold work. For example, metals show an increase in hardness, an increase in critical shear stress, a decrease in ductility, and an increase in electrical resistivity as a result of irradiation. However, since recent experiments have shown that there are many properties of metals and alloys which behave dissimilarly with respect to cold work and radiation damage, the analogy is no longer considered useful.

Additional reasons for discontinuing the analogy between cold work and radiation damage deal with the types of lattice defects introduced by the two different mechanisms. Cold work involves the motion and multiplication of dislocations, while irradiation is not expected to introduce such organized lattice disturbances. Differences in effects on properties are expected on this basis. X-ray line broadening, for example, which is appreciable in metals cold-worked at 20°C and is successfully interpreted on the basis of dislocation theory, has not been observed in irradiated metals.

Recent theoretical work by F. Seitz indicates that some vacancies are produced by cold work. Interstitials may also be introduced by plastic deformation, but it requires higher energies to produce interstitials than vacancies, at least in metals. One might therefore expect the number of interstitials created by cold work to be rather small. No such restrictions apply to the production of interstitials by fast particle irradiation. Significant differences in the annealing behavior of cold-worked and irradiated materials have been observed, which may be explicable by such considerations. In summary, one may learn more from a critical study of the differences rather than the similarities of irradiation effects and the effects of cold work.

5·9 Radiation-damage effects on metals

In general, it has been shown by many investigators that most metals and alloys become harder and stronger as a result of neutron bombardment, particularly if the temperature is sufficiently low. The most striking effect on mechanical properties is the increase of the critical shear stress of copper single crystals from the unirradiated value of 280 to 10,600 psi. Similar results were obtained on iron and zinc single crystals.

Studies on aluminum and copper single crystals after bombardment at 80°K (−193°C) revealed that the effects in aluminum annealed out before reaching room temperature, while temperatures of 300°C were required to anneal out the increase in critical shear stress in copper. This result and other studies of many different metals indicate strongly that there is a relationship between diffusion and the temperature stability of radiation effects in pure metals, radiation effects being greater the lower the rates of self-diffusion (diffusion of x atoms in metal x) at a given temperature. Thus, these experiments furnish indirect evidence for vacancies and interstitials, since diffusion can annihilate these defects in accordance with the experimental observations. Furthermore, the creation of dislocations seems unlikely because of the particular temperature dependence of the radiation lattice defects.

Besides mechanical behavior, the effects of neutron irradiation on a number of other properties of pure metals have been studied. The electrical resistivity of copper increased by 0.25 per cent during pile irradiation at 27°C and by over 30 per cent by irradiation at 80°K (−163°C). Other metals show similar effects, in approximate relation to the self-diffusion rate at the temperature of the irradiation. The elastic constants of copper have been measured before and after irradiation, and large increases have been reported.

5·10   Radiation-damage effects on precipitation-hardening alloys

The effects of neutron bombardment on the physical properties of precipitation-hardening alloys have not been studied extensively. Work at Oak Ridge has shown that neutron irradiation can cause precipitation in a metastable supersaturated solid solution of Cu-Be. The effect of room-temperature (20°C) irradiation on the Cu-Be alloys was found to be similar to the effects of aging at approximately 100°C. This result was attributed to a greater ease in local diffusion due to the excess number of vacancies and interstitials introduced during irradiation. G. T. Murray and W. E. Taylor, who carried out much of this work at Oak Ridge, irradiated many other precipitation-hardening alloys at room temperature but did not observe any property changes. They suggested that their results might be explained by one or more of the following causes:

1. The base metal may not be heavy enough for large numbers of displacements per incident neutron.
2. The activation energy for diffusion may be too low. In other words, the diffusion rate at low temperatures (near room temperature) may be too low to permit movement of vacancies or interstitials.
3. The size of the stable precipitate particle may be too large.
4. The precipitate particles may be too small to affect the measured physical properties.

Further experiments on Cu-Be alloys showed that a much smaller effect due to irradiation was observed after bombardment at 77°K (−196°C). However, the Cu-Be alloys gave evidence of precipitation after being warmed to room temperature. This result substantiates the arguments for precipitation caused by the diffusion of the excess vacancies and interstitials introduced by neutron irradiation.

The effects of radiation damage on precipitation-hardening alloys are of great importance in determining the overall usefulness of this class of alloys in reactors. The results of the experiments on Cu-Be alloys show that irradiation can cause precipitation from supersaturated solid solutions. Such an effect on an alloy heat-treated to maximum properties could lead to overaging and a decrease of strength and hardness. Experiments on Cu-Fe alloys by J. Denny of North American Aviation and A. Boltax at MIT and Nuclear Metals, Inc., indicate that neutron irradiation can cause re-solution of precipitates present at the start of the irradiation. Re-solution of precipitates would also result in a decrease of strength and hardness in heat-treated precipitation-hardening alloys. Therefore, as a result of these experiments, reactor designers must consider the phenomena of radiation-induced re-solution or overaging in their design considerations and materials selection.

5·11   Radiation-damage effects on order-disorder alloys

Alloys which can develop a superlattice structure (long-range ordering) behave similarly to precipitation-hardening alloys. Two conditions can exist at low temperatures, one in which the alloy has been slowly cooled and is an ordered state, the other in which the metal, having been quenched from the disordered region, is disordered. Radiation damage would be expected to show two effects. The interstitials and vacancies formed may fall back into different sites, thus tending to disorder a completely ordered alloy. On the other hand, there will be a tendency for them to fall back into sites with short-range order (local ordering of atoms), thus possibly ordering a disordered alloy.

The first experiment of this kind was reported by S. Siegel on the Cu_3Au alloy. He found that the ordered alloy became disordered as a result of fast neutron irradiation, while the disordered alloy did not order to any observable extent. Later experiments, however, demonstrated that both ordering and disordering could be caused by irradiation.

When calculations of the number of atoms disordered per incident neutron are made, the results cannot be explained by displacement collisions (production of vacancies and interstitials) alone. The disordering of ordered Cu_3Au and Ni_3Mn by bombardment at room temperature lends some credibility to the concept of a thermal-spike mechanism, since displacements by collision would not appear to disorder these samples as rapidly as observed. L. Aronin has estimated that one neutron of 1 mev effects the replacement of 5,000 atoms in the disordering process, that is, a region about 70 Å in diameter.

5·12   Radiation-damage effects on nonmetals

The effects of neutron irradiation on nonmetals such as ionic compounds [e.g., quartz (SiO_2) and lithium fluoride (LiF)], ceramics, and organic materials are considerably more drastic than the effects on metals. The reasons for this are complex and varied for each of these materials, but a few general statements can be made.

In addition to the displacement effects observed in metals, insulators (ionic compounds) will show damage effects due to ionization. As mentioned previously, the energy of the primary recoil atom is dissipated by elastic collisions, ionization collisions, and ionization of electrons. In any good conductor (most metals) the ionization effects will disappear very quickly, resulting in the heating of the material. However, in an insulator, the electrons liberated by ionization may be trapped at various lattice imperfections, permanent changes thus resulting. For example, the room-temperature thermal conductivity of quartz has often been observed to drop by a factor of 10 after moderate irradiation.

Another important factor which determines the overall effect of radiation damage is the ability of the material to flow plastically and thus relieve excessive internal stresses. LiF crystals will shatter after moderate exposures owing to the strain induced by the release of helium and tritium from the (n,α) reactor in Li^6. Metal ceramics will show considerably larger radiation-damage effects than metals at low temperatures, because they cannot relieve internal stresses by plastic flow. At elevated temperatures, ceramic materials develop plastic flow, and the change in properties under high-temperature irradiation may be considerably less.

Those nonmetals which are characterized by covalent bonding appear to be most susceptible to dimensional changes as a result of neutron irradiation. Lattice expansion has been reported in both graphite and diamond after irradiation. Plastics show drastic effects, such as decomposition and brittleness, when subjected to irradiation. In general, organic materials built up from benzene rings appear to be somewhat more irradiation-stable than other organics.

5·13   Effects of radiation damage on uranium

A fuel material is subjected to both fast neutron and fission fragment damage, and thus the damage observed is unusually severe. Most of the damage in uranium appears in the form of embrittlement, distortion, growth and swelling. This problem has received a great deal of attention since 1960 but, at present, is still unsolved. However, much has been accomplished, both from an experimental and a theoretical point of view. In the next two sections, we shall examine some of the details of this problem. It should be added that, from a scientific viewpoint, this is one of the most complex and least understood of all materials problems and, furthermore, is possibly the single most important problem standing in the way of economic power reactors.

The behavior of fissionable materials is usually examined as a function of burnup and temperature history. The units of burnup employed depend on the ordinary usage for a given fuel system. For example, the unit atomic per cent burnup (relative to the total number of atoms in the fuel, not including the cladding) is used for uranium, plutonium, and their alloys, whereas the unit megawatt-days per ton (MWD/T) is used for UO_2, for ThO_2-UO_2, and also for uranium and uranium alloys. Recently, a great deal of interest has developed in the burnup unit fissions per cubic centimeter, which is useful for comparing fuels on a volume basis. Since the engineers and designers in various groups use different burnup units, it is important for the student to learn how to use them all. The conversion factors for the various burnup units are given below for natural uranium,

$$1 \text{ atomic } \% \text{ burnup} = 8{,}200 \text{ MWD/T} = 4.8 \times 10^{20} \text{ fissions/cm}^3$$

(assuming 1 per cent of all atoms fissioned, with 192 mev liberated per atom fissioned). The details of this conversion should provide an interesting exercise for the student.

5·14 Anisotropic growth

Alpha uranium, which is stable up to 663°C, is orthorhombic in structure and has markedly anisotropic properties (i.e., its properties within a single-crystal or polycrystalline metal with a high degree of preferred orientation are dependent upon crystalline direction). For example, the coefficients of thermal expansion vary from slightly negative values [in the (010) direction] to large positive values [in the (100) and (001) directions] as a function of crystal orientation. Other examples of important properties which are anisotropic are the diffusion coefficients and mechanical properties. A further complication is that uranium exists in three allotropic modifications in the temperature range from room temperature to 765°C. Alpha uranium transforms to the beta phase at 663°C, and it in turn transforms to the gamma phase at 765°C.

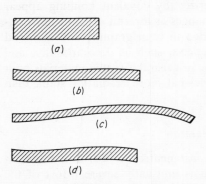

Fig. 5·6 Sketches of uranium growth and distortion caused by thermal cycling and neutron irradiations. (a) Original uranium cylinder after rolling at 300°C and machining; (b) after 1,300 cycles between 50 and 500°C; (c) after 3,000 cycles between 50 and 500°C; (d) after 0.1 per cent burnup.

Based on this combination of properties, it was not surprising that appropriate combinations of heat-treatment were observed to cause growth and distortion even in the absence of irradiation. Growth of uranium as a function of *thermal cycling* was first observed at Battelle Memorial Institute. Thermal cycling is the name given to a heat-treatment which consists in alternating the temperature of a specimen between two fixed values. Growth and distortion of uranium were first reported in massive uranium as a result of neutron irradiation by the Hanford Laboratories. The sketches of Fig. 5·6 illustrate the order of magnitude of the growth and distortion caused by thermal cycling and neutron irradiation.

The resemblance between thermal cycling and irradiation distortion led to much study of thermal cycling at a number of USAEC laboratories. The results indicated that preferred orientation, grain size, microstructure, and time-temperature relationships all play an important role in determining the amount of distortion as a result of thermal cycling. The kind and extent of distortion obtained under irradiation are related to the size and preferred orientation of the metal grains, fabrication techniques, and subsequent heat-treatment.

Several mechanisms have been proposed to explain the growth caused by thermal cycling and irradiation. Perhaps the best known is the "ratchet" mechanism proposed by the staff at Knolls Atomic Power Laboratory. The suggested mechanism depends upon the ease of grain boundary flow vs. crystallographic slip as a function of temperature, coupled with the anisotropy of the thermal expansion coefficients of uranium to explain thermal cycling in the absence of neutron irradiation. Efforts to apply this mechanism to distortion during irradiation have not been entirely successful, the principal discrepancy being that single-crystal uranium samples are unaffected during thermal cycling but are greatly affected by irradiation.

Recent work by the British has led them to suggest that the stresses set up in the region of the fission spike lead to anisotropic deformation, causing a change in shape of a crystal which can be related to its crystallographic orientation. Each fission event can be divided into a heating and a cooling period. In the heating period, the material surrounding the fission spike is in compression, and it is postulated that plastic deformation occurs preferentially in the [010] direction of the crystal because of the low compression yield stress in this direction. During subsequent cooling, the compressive stresses are replaced by tensile stresses, and because of the change in yield behavior with the direction of stress, plastic deformation occurs preferentially normal to the [010] direction. The net result of these processes is that a crystal changes shape on irradiation, tending to grow in the [010] direction.

In a randomly oriented polycrystalline aggregate, such changes in shape require associated deformations in the bulk of the material if no internal voids are formed. At the surface of such a specimen, changes in shape can take place more freely, and crystals having their [010] directions normal (or nearly so) to the surface may be expected to grow outward and cause irregularities on a previously smooth surface. This phenomenon is generally referred to as *wrinkling*. In material which is not randomly oriented but has a marked texture, it is to be expected that the change in crystal shape due to irradiation will cause a change in shape of the whole sample. This has also been observed and has been termed *growth*.

Reduction of wrinkling and growth in polycrystalline material can clearly be effected by obtaining a randomly oriented fine-grain crystal structure. This has been shown to be achieved by rapidly cooling through the alpha-beta transformation (called beta quenching). The efficiency of this treatment has been found to depend markedly on the purity of the material, and it is also sensitive to mass effects, so that results from small laboratory specimens require considerable development to translate them into effective practical processes.

The last theory we shall examine is that proposed by L. L. Seigle and A. J. Opinsky. They have postulated that shape changes occur by anisotropic diffusion of interstitials or vacancies. As the details of this theory are rather complicated, the description will be only schematic. If vacancies preferentially diffused to dislocations in particular directions, the extra planes would gradually disappear and contraction of the lattice would take place. At the same time, extra planes could be formed by directional diffusion of interstitial atoms, giving the observed shape changes, such as growth and wrinkling.

5·15 Swelling

In the fission of a uranium atom two fission-product atoms are released into the lattice. These impurity atoms give rise to an inherent volume increase, because two atoms have been created for each uranium atom fissioned. A density change of approximately 3.4 per cent per 1 per cent of the uranium atoms fissioned would be expected because of this phenomenon. This behavior would be approached by real fuel alloys operating in a temperature region where there is no diffusion of the fission-product atoms. Since approximately 10 per cent of the fission-product atoms are the rare gases Xe and Kr, larger density changes could occur if the Xe and Kr atoms were sufficiently mobile to collect and form gas pockets in the metal-fuel system.

Gas pockets do form in irradiated fuel materials. Density changes of 10 to 100 per cent per atomic per cent burnup have been reported. Such behavior is called swelling and is to be distinguished from impurity atom effects, which are a natural consequence of the fission reaction, whereas swelling can occur only when the fission products are sufficiently mobile to diffuse and precipitate as gas bubbles.

Several techniques, most of them under current study, have been suggested as ways of coping with the swelling problem in fissionable material. A partial list follows:

1. Application of external pressure by means of a strong cladding.
2. The use of dispersion-type fuel elements. In a dispersion-type fuel element, the fissionable phase is surrounded by a matrix of nonfissionable material, which serves to restrain the swelling which occurs in the fissionable phase.
3. The use of uranium compounds such as uranium dioxide and uranium carbide as the fuel material. Uranium compounds have a high melting point and low density, which contributes to the improved swelling resistance of these materials.

5·16   Effects of radiation damage on thorium

Thorium is fcc and up to the melting point exhibits no allotropic trans-formations. The isotropic nature of thorium as contrasted with uranium would indicate that one would expect less radiation damage in thorium. In general, the effects that have been observed thus far do indicate its excellent radiation stability. Experiments reported at the Geneva Conference of 1955 show that hot-rolled thorium is dimensionally stable under irradiation (maximum exposure 10^{21} nvt).

5·17   Effects of radiation damage on uranium alloys

The severe radiation damage observed in metallic uranium has led to studies of various alloys such as uranium-aluminum, uranium-beryllium, uranium-chromium, uranium-molybdenum, and uranium-zirconium. Alloy-ing of uranium is aimed at producing a fuel-element material that is dimen-sionally stable during use in a reactor. The effects of alloying on the structure of uranium are varied. For example, a uranium compound may be dispersed in a matrix of a soft metal, or a two-phase structure similar to a precipitation-hardened microstructure may be formed, or a high-temperature isotropic phase may be retained by fast cooling from the high temperature.

One of the first uranium-alloy systems to be studied was that of uranium and aluminum containing up to 30 per cent uranium by weight. As shown in the phase diagram of Fig. 5·7, the phases present at room temperature in all alloys up to 30 weight per cent uranium are the compound UAl_4 and essentially pure aluminum. The microstructure of these alloys is such that the aluminum furnishes the continuous matrix, and therefore the uranium can be considered as dispersed in the aluminum. There is a eutectic reaction occurring at approximately 13 weight per cent uranium, so that alloys near this composition show the highest degree of uniform dispersion.

The results of irradiation of alloys containing 5, 15, and 30 weight per cent uranium were very encouraging. No changes as great as 1 per cent were observed in the original dimensions. Density measurements showed a maximum density decrease of 3 per cent at an 0.05 per cent burnup. The irradiation did, however, cause a large decrease of the ductility, but annealing experiments showed that the ductility would increase considerably after a short anneal (1 hr) at 400°C.

In summary, the uranium-aluminum alloys studied showed no serious changes even after 1 atom in 1,000 had fissioned, while, in pure uranium, severe damage was observed after 1 atom in 10,000 had fissioned. This improved behavior can be attributed to the dispersion of the uranium in a metal well suited to absorb the damage from the fast neutrons and the fissioning process.

An alloy of uranium containing 9 per cent molybdenum has shown several interesting properties. At the 1955 Geneva Conference, the Russians reported

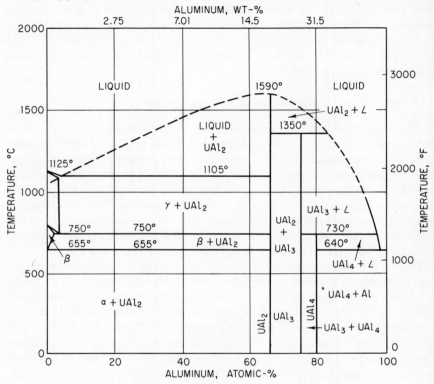

Fig. 5·7   The uranium-aluminum phase diagram.

that the irradiation of uranium is analogous to thermal treatment, as confirmed by the behavior of a uranium–9 per cent molybdenum alloy. The equilibrium diagram of Fig. 5·8 predicts that the phases present at room temperature in the 9 per cent alloy are α uranium plus an intermetallic compound epsilon (ϵ). However, if the alloy is not slowly cooled from above 600°C, the bcc gamma (γ) phase is retained at room temperature. The solid-state eutectoid transformation of the γ phase to the $\alpha + \epsilon$ phases is very sluggish.

The effects of neutron irradiation on the γ phase and on the $\alpha + \epsilon$ mixture of phases were of particular interest. The experiments showed that, while the specimens of the homogeneous γ phase preserved their structure under irradiation, the specimens of the heterogeneous alloy were found to have become partly or completely homogeneous γ phase. These results were confirmed by recent experiments carried out jointly by several USAEC laboratories.

The Russian reports offered no explanation of the radiation-induced phase change. The experiments carried out in the United States were reported by

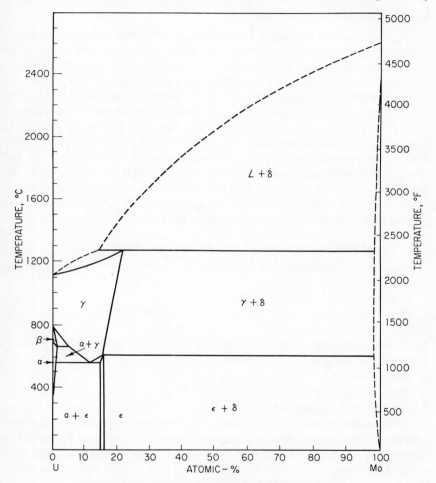

Fig. 5·8   The uranium-molybdenum phase diagram.

the Westinghouse Atomic Power Group, which suggested that the phase change may be attributed to displacement spikes which permit enough atomic movement to homogenize these alloys. The experimental results were qualitatively explicable by using a model based on Brinkman's displacement-spike hypothesis. It is interesting to note that the irradiation-stable structure is the bcc form, which is the desirable structure to minimize radiation instability, as it is more isotropic than the orthorhombic structure.

PART 3. CORROSION

5·18   Liquid-metal corrosion

The major factors limiting the design and construction of high-temperature nuclear reactors are the properties of suitable coolant containers and structural materials. These properties include mechanical strength, dimensional stability, heat-transfer characteristics, the ability to be fabricated into complex leakproof piping systems, and, of particular interest in this section, the ability to withstand attack by liquid metals. This attack will be discussed in its more general sense and will be termed *corrosion*. The broad problem of corrosion by liquids other than liquid metals will not be considered here. This latter problem has received extensive attention during the last half century, and many textbooks have been written which adequately cover this field. The present discussion will be limited to liquid-metal corrosion, as this problem will become increasingly important as we progress toward the goal of high-temperature economic power reactors.

5·19   Types and mechanisms of liquid-metal corrosion

Liquid-metal corrosion may take place by several simple types of mechanisms:
1. Solution attack
2. Direct alloying
3. Intergranular penetration
4. Corrosion by contaminants
5. Corrosion erosion
6. Self-welding
7. Thermal-gradient metal transfer
8. Concentration-gradient metal transfer

The first is due to dissolving of the container material in the liquid metal. Direct alloying is the interaction between liquid and solid to form surface films or typical diffusion layers of intermetallic compounds and solid solutions. These may form a loosely adherent scale or, if held tightly, may serve as a barrier to slow down additional diffusion.

Selective reaction of the liquid metal with minor constituents of the solid may result in intergranular penetration or in the depletion of a dissolved component of the solid. Selective grain-boundary attack can drastically alter the physical properties of a material without appreciably changing its weight or appearance. This type of attack is often accelerated by the application of stress to the solid during exposure to the liquid metal.

Corrosion by contaminants in the liquid metal rather than by the liquid metal itself is another form of corrosion. Contaminants may include oxygen, nitrogen, or carbon, dissolved in the liquid or present as part of a compound in suspension. For example, in oxygen-contaminated systems, the container metal can become coated with a layer of its own oxide, provided that its oxide is more stable than the oxide formed by the liquid metal. This reaction is represented by the following schematic chemical equation:

$$M_{\text{liquid}} O + M_{\text{container}} \rightarrow M_{\text{container}} O + M_{\text{liquid}} \tag{5·3}$$

If the liquid-metal oxide were more stable than the container oxide, no oxide coating would form on the container. The degree of stability of metal oxides is given in terms of the energy of formation (actually these values are known as the free energy of formation; however, the explanation of this concept is beyond the scope of this discussion). Table 5·4 shows typical values of the

Table 5·4   Energy of formation of metal oxides

Sample reaction: $M_{\text{solid}} + \frac{1}{2}O_{2(\text{gas})} \rightarrow MO_{\text{solid}}$

Metal	Metal oxide	Energy of formation* at room temperature, kcal
Li	Li$_2$O	−134
Na	Na$_2$O	− 91
K	K$_2$O	− 86
Zn	ZnO	− 76
Cd	CdO	− 54
Pb	PbO	− 45
Ag	Ag$_2$O	− 3
Au	Au$_2$O$_3$	+ 39

* The more negative the value of the energy of formation, the more stable the oxide. Li$_2$O is the most stable oxide given in the table. A positive energy-of-formation value indicates that oxidation does not take place, as in the case of gold (Au).

energy of formation of a few oxides. From the energies of formation one can generally determine whether or not an oxide can be expected to form. For example, from Table 5·4 one would predict the following reactions:

$$2\text{Li}_{\text{liquid}} + \text{PbO}_{\text{coating}} \rightarrow \text{Li}_2\text{O}_{\text{solid in liquid}} + \text{Pb}_{\text{solid}} \tag{5·4}$$

$$\text{PbO}_{\text{solid in liquid}} + \text{Zn}_{\text{solid}} \rightarrow \text{ZnO}_{\text{coating}} + \text{Pb}_{\text{liquid}} \tag{5·5}$$

Equation (5·4) represents the case of liquid lithium flowing through a lead pipe which is coated with lead oxide. Equation (5·5) likewise, represents the case of liquid lead containing some oxygen in the form of the suspension PbO flowing through a zinc pipe. In the first, the oxide coating is removed by the liquid metal, while, in the second, a coating of oxide is formed on the zinc surface.

As in the case of alloy layers, the oxide layer is sometimes tenacious and adherent and acts as a diffusion barrier, in which case it inhibits further attack. On the other hand, it may be nonadherent, in which case drastic weight loss ensues, especially under dynamic conditions.

Another type of attack, related to corrosion but more appropriately termed corrosion erosion, results from mechanical attack by flowing or turbulent liquid metal and involves the removal of a scale, abrasion by suspended particles, or, in extreme cases, *cavitation*. Cavitation is the damage caused by the formation and collapse of cavities in a moving liquid at the solid-liquid interface.

One of the most serious manifestations of liquid-metal behavior, in which actual corrosion plays only a small part at the most, is that which results in *diffusion bonding*, or *welding*, of solid metal surfaces to each other. Such self-welding is facilitated by contact with an alkali metal (which may clean the surface by dissolving oxides or other coatings) and is intensified if the metals are held together under pressure. This effect becomes increasingly serious with increasing temperature. The possibility of its interfering with the operation of valves, pumps, and flanged joints may well be a major factor in determining the upper temperature limit of operation of a liquid-metal heat-transfer system.

Of particular interest in liquid-metal systems is a type of mass transport, thermal-gradient transfer, which is due to the coexistence of a temperature differential and an appreciable thermal coefficient of solubility. Even though the actual solubility may be quite low, large amounts of a solid component may be dissolved from the zone of higher solubility and precipitated in the zone of lower solubility. This continuous removal of the dissolved component from the system accelerates corrosion attack in some areas, while the precipitated material accumulates in other parts of the system and may eventually cause plugging of flow channels. The rate of transfer of container material from one zone to another is observed to be greater for some systems, e.g. iron in mercury, than for others, e.g., iron in sodium, even though the reverse of this would be expected from consideration of equilibrium solubilities alone. Research workers in this field have attributed the phenomenon to the relative influence of solution rate as opposed to diffusion rate as the corrosion-limiting factor. Attack on steels by mercury is inhibited by the addition of titanium or zirconium. These inhibitors were found after very extensive and expensive laboratory investigations, and although possible explanations have been offered, the mechanism by which they act is unknown. Certain impurities, especially oxygen, may accelerate solution attack and the thermal-gradient transfer effect. Where oxygen is known to accelerate corrosion by liquid metals, inhibitors are being sought which will tie up the oxygen as an insoluble oxide.

Another type of mass transfer of material from one part to another of a multimetallic system is concentration-gradient transfer, which may be encountered even in an isothermal (uniform temperature) system. Instead of the dissolving of one metal stopping when the liquid is saturated, the equilibrium may be upset by the presence of a second metal with which the dissolved metal can alloy to form solid solutions or compounds by precipitation in a surface diffusion layer. An alloy layer thus formed may act as a diffusion barrier, causing the mass-transfer rate to decrease with time.

5·20 Some data on the resistance of metals to corrosion by liquid metals

The evaluation of materials according to their corrosion behavior in liquid metals is summarized in Fig. 5·9. The ratings are based on the criteria of Table 5·5. While the charts are based on tabulated data, they represent for

Table 5·5

Rating	Rate of attack, mils/year
Good	Less than 1
Limited	Greater than 1, less than 10
Poor	Greater than 10

the most part only a limited number of relatively small-scale laboratory tests. It should not be assumed that the same numerical rates would apply in larger-scale heat-transfer systems or that these rates are independent of time or the surface-to-volume ratio of the experiment. The bar charts have the principal value that they can be used for the elimination of many materials from consideration and for the selection of the more promising materials for further evaluation of physical and mechanical properties, as well as for direct experimental investigation in simulated service environments.

PART 4. THERMAL PROBLEMS

5·21 Thermal deformation

In all types of equipment which are designed for the production of heat, there exists a problem of selecting materials which are capable of withstanding sharp thermal gradients and of retaining their original shape, even after prolonged use at high temperatures. These problems have confronted boiler and furnace designers for years and now must be considered by reactor designers in the use of materials for reactor components. We shall examine two thermal problems in particular: (1) *thermal stress* and (2) *creep*, or *flow of metals* below their normal yield strength.

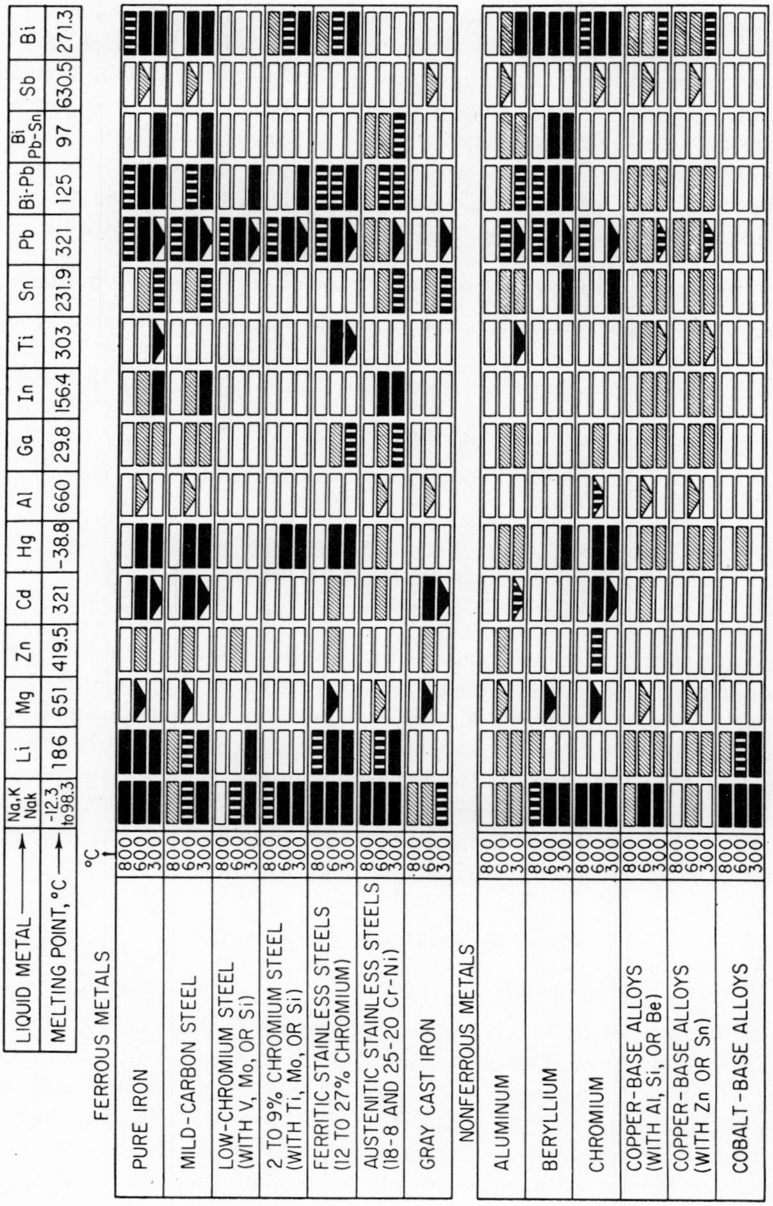
LIQUID METAL
MELTING POINT, °C
Na,K Nak
-12.3 to 98.3
Li
186
Mg
651
Zn
419.5
Cd
321
Hg
-38.8
Al
660
Ga
29.8
In
156.4
Tl
303
Sn
231.9
Pb
321
Bi-Pb
125
Bi Pb-Sn
97
Sb
630.5
Bi
271.3
°C
FERROUS METALS
PURE IRON
MILD-CARBON STEEL
LOW-CHROMIUM STEEL (WITH V, Mo, OR Si)
2 TO 9% CHROMIUM STEEL (WITH Ti, Mo, OR Si)
FERRITIC STAINLESS STEELS (12 TO 27% CHROMIUM)
AUSTENITIC STAINLESS STEELS (18-8 AND 25-20 Cr-Ni)
GRAY CAST IRON
NONFERROUS METALS
ALUMINUM
BERYLLIUM
CHROMIUM
COPPER-BASE ALLOYS (WITH Al, Si, OR Be)
COPPER-BASE ALLOYS (WITH Zn OR Sn)
COBALT-BASE ALLOYS
800
600
300

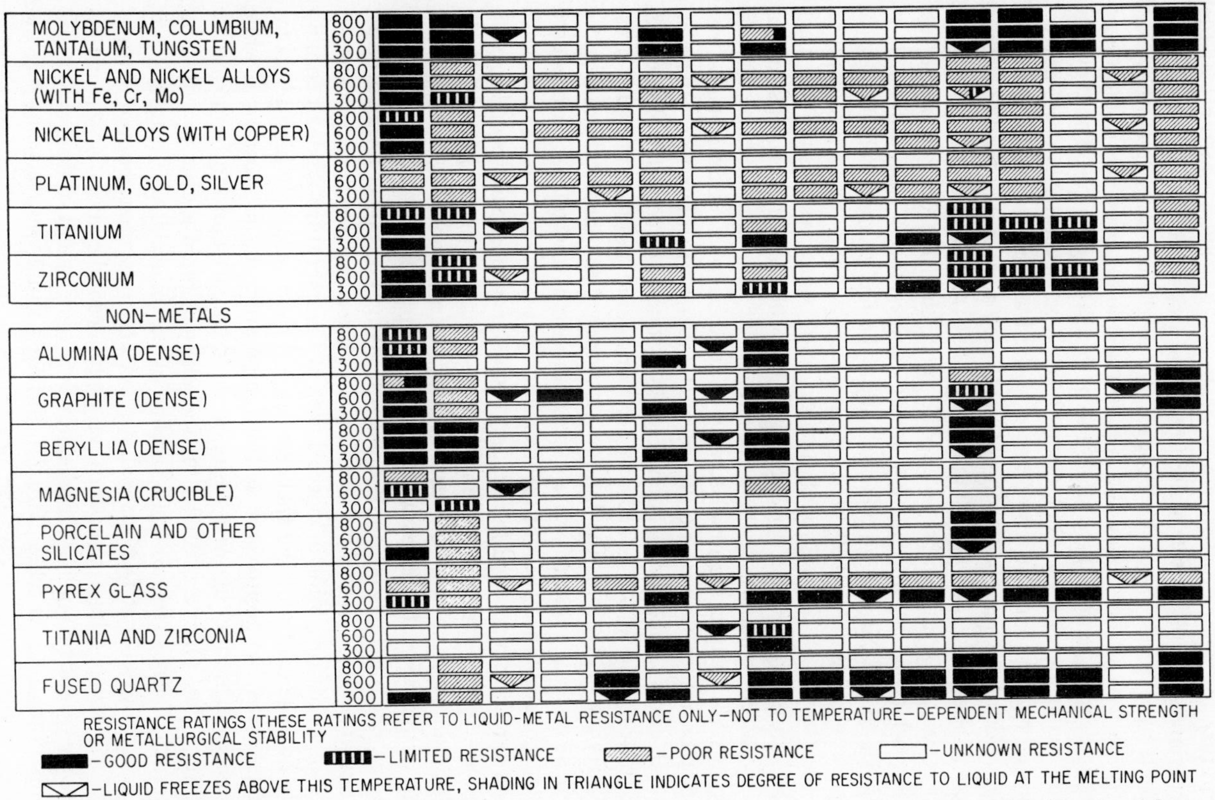

MOLYBDENUM, COLUMBIUM, TANTALUM, TUNGSTEN	800 600 300
NICKEL AND NICKEL ALLOYS (WITH Fe, Cr, Mo)	800 600 300
NICKEL ALLOYS (WITH COPPER)	800 600 300
PLATINUM, GOLD, SILVER	800 600 300
TITANIUM	800 600 300
ZIRCONIUM	800 600 300

NON-METALS

ALUMINA (DENSE)	800 600 300
GRAPHITE (DENSE)	800 600 300
BERYLLIA (DENSE)	800 600 300
MAGNESIA (CRUCIBLE)	800 600 300
PORCELAIN AND OTHER SILICATES	800 600 300
PYREX GLASS	800 600 300
TITANIA AND ZIRCONIA	800 600 300
FUSED QUARTZ	800 600 300

RESISTANCE RATINGS (THESE RATINGS REFER TO LIQUID-METAL RESISTANCE ONLY—NOT TO TEMPERATURE—DEPENDENT MECHANICAL STRENGTH OR METALLURGICAL STABILITY

■—GOOD RESISTANCE   ▨—LIMITED RESISTANCE   ▧—POOR RESISTANCE   ▯—UNKNOWN RESISTANCE

▨—LIQUID FREEZES ABOVE THIS TEMPERATURE, SHADING IN TRIANGLE INDICATES DEGREE OF RESISTANCE TO LIQUID AT THE MELTING POINT

Fig. 5.9 Condensed summary of resistance of materials to liquid metals at 300, 600, and 800°C. (Reprinted from "Liquid Metals Handbook.")

5·22   Thermal stress

The basic principle of heat flow is that heat flows from regions of high temperature to regions of lower temperature. The central portions of fuel elements are generally at the highest temperatures reached in a reactor. The coolest temperatures, naturally enough, are in the coolant; thus, there is a continuous flow of heat from the fuel to the coolant, and there is a corresponding drop in temperature along the path of the heat flow. The temperature at the center of the fuel element may be several hundred degrees centigrade higher than the temperature at the fuel-element surface, and the respective distances may be less than $\frac{1}{2}$ in. Since most metals tend to expand as their temperature increases, one realizes that the center of a fuel element, which is relatively hot, wants to expand more than the outside of the fuel element, which is at a relatively lower temperature. This difference in tendencies to expand sets up a stress within the fuel element which, if great enough, can cause the fuel element to distort out of shape or even to rupture.

In the design of conventional furnace and boiler systems, it has been possible, to a large extent, to overcome the problems of thermal stress by carefully selecting materials which are capable of withstanding large thermal stresses without rupturing or distorting or materials which do not give rise to large thermal stresses, even under severe temperature differences. However, as we well know, a nuclear fuel comprises only one of three or four materials, which, unfortunately, are not ideal when considered from the thermal-stress point of view. Therefore, the limitations imposed by the establishment of thermal stresses in nuclear fuels are quite important and must be considered in the design and fabrication of fuel elements. In practice, it has been possible through carefully controlled fabrication procedures to impart special crystallographic structures to fuel elements so that either less thermal stress is built up within an operating fuel element or else the fuel element is better able to withstand the thermal stresses which are built up within it.

When a reactor starts up, shuts down, or changes its power level abruptly, a particular type of thermal stress, known as *thermal shock*, may arise. A reactor in steady operation achieves a type of "temperature-gradient equilibrium"; i.e., the center of the fuel reaches a certain relatively high temperature and remains at approximately this temperature during operation, while the outside of the fuel likewise assumes an essentially constant temperature at a lower value. One reason why the fuel element often is able to withstand its internal thermal stresses is the ability of the material to "readjust," or flow plastically, so that the internal stresses are somewhat relieved. This is particularly true at high temperature, when even the most brittle metals show some plastic behavior. There is, however, an element of time associated with this type of stress relief. Thus, if, during a shutdown, for example, the fuel were cooled too rapidly, the new thermal stresses which are created during rapid cooling might not be self-relieved fast enough. In such cases, there may be distortion or rupture of the fuel element (or any other material in the reactor in which a thermal stress could occur). Thus, in the operation of a reactor, it may be desirable to control the rate of start-up, shutdown, or power-level change, so that the possibility of thermal-shock damage will be held to a minimum.

In the case of ceramic fuel elements, the problems of thermal stress, and particularly thermal shock, are of great importance. Ceramics, in general, possess little ductility compared with metals and are usually very brittle. Thus, it is conceivable that internal thermal stresses within a ceramic element could actually shatter the element, in the way in which certain types of glass shatter when subjected to large thermal stresses. The use of *cermets* (combinations of metals and ceramics) may be helpful in reducing the deleterious effects of thermal shock in a ceramic type of fuel. As in the case of metallic fuels, the method of fabrication plays an important part in the ability of a ceramic fuel to withstand thermal shock.

5·23 Creep

Creep may be defined as the time-dependent portion of the strain imposed upon a metal by the application of a stress. From a phenomenological viewpoint, *creep* generally appears as the elongation or flow of metals held for long periods of time at stresses lower than their normal yield strength. Creep is not a unique property of metals but is a general phenomenon observed in essentially all solids, e.g., glasses, plastics, and even concretes.

The characteristic form of a creep curve is that of curve *A* in Fig. 5·10, while curves *B* and *C* are typical variants of it. There are four main features of a creep curve, as indicated in Fig. 5·10:

1. The initial "instantaneous" strain, partly elastic (*a*)
2. The period of decelerating flow (*b*)
3. The period of constant minimum rate of flow (*c*)
4. The period of accelerating flow, which ends in fracture (*d*)

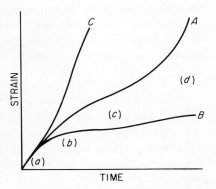

Fig. 5·10 Characteristic form of creep curves. (A) Standard form; (B) at low stresses and temperatures; (C) at high stresses and temperatures.

There is a widely held but mistaken notion that creep is purely a high-temperature phenomenon. The error arose because the type of creep that worries the engineer is that obtained at high temperatures, where a metal has a short working life and flows too fast for its dimensions to remain tolerably stable at any stage. Until 1950 the type of creep shown in curve *B*, (Fig. 5·10) which can be obtained even at the very lowest temperatures, has aroused little interest because it is a transient flow which soon dies away almost completely, so that the dimensions of the specimen become stable and the applied load can be supported safely for an indefinitely long time. In recent years, however, the development of the gas turbine has quickened interest in all aspects of creep.

The factors which determine the extent of creep in a given material are the magnitude of the load and the temperature. The rate of elongation may decrease with time, remain constant, or increase with time (see Fig. 5·10), depending upon the mechanisms operating during creep. These mechanisms can be examined at two different levels. Advanced theories of creep deal with the fundamental atomic movements which cause the overall extension of a material, while macroscopic theories based on structure and hardness changes form the basis for engineering theories of creep. A discussion of either of these theories is beyond the scope of this course.

The choice of material is probably the most important method of controlling creep. In applications where high creep strength is the most important consideration, attention must be given to iron, nickel, or cobalt-base alloys containing considerable amounts of chromium, tungsten, molybdenum, niobium, titanium, and zirconium. Chromium, in particular, is essential for corrosion resistance to air at high temperatures. Cast alloys are useful when creep resistance is the chief requirement; but where dynamic stresses at high temperatures are encountered, wrought (worked) alloys are preferred. Wrought age-hardenable alloys are successful if they are not used at temperatures that promote overaging.

PROBLEMS

Note the correct statements; there may be none, one, or more than one.

1. The main differences in materials problems between solid-fuel and liquid-fuel reactors are (a) the temperature of operation; (b) the effect of radiation damage; (c) the problem of heat transfer; (d) the reactor power levels.

2. Radiation damage in a solid may arise in several ways, among which are (a) through collisions with fission fragments; (b) through collisions with thermal neutrons; (c) through ionization effects; (d) by capture of thermal neutrons.

3. The physical consequences of radiation damage result in the introduction of defects into the lattice, such .as (a) vacancies; (b) interstitials; (c) dislocations; (d) grain boundaries.

4. Vacancies and interstitials may affect the properties of solids, such as (a) grain size; (b) hardness; (c) electrical resistivity; (d) critical shear stress.

5. A thermal spike refers to (a) rapid heating of a small mass of material;

(b) a hot nail; (c) localized heating resulting from the dissipation of energy of a fast particle; (d) the temperature of the fuel element during irradiation.

6. The similarities between radiation damage and cold work involve (a) the effects on hardness, strength, and ductility; (d) the effects on microstructure; (c) the effects of X-ray line broadening; (d) the effects on dislocation density (the number of dislocation lines per square centimeter).

7. The magnitude of the effects of radiation damage on metals is dependent upon (a) the volume of the metal; (b) the temperature of irradiation; (c) the rate of diffusion in a particular metal; (d) the neutron flux.

8. The effects of neutron irradiation on precipitation-hardening alloys are (a) precipitation from supersaturated solid solutions if the diffusion rate of vacancies and interstitials is high at the temperature of the irradiation; (b) general recrystallization; (c) re-solution of small precipitate particles; (d) general hardening.

9. The factors which contribute to uranium's particular sensitivity to radiation damage and thermal cycling are (a) high melting point; (b) anisotropic physical properties, (c) three allotropic modifications; (d) high density.

10. The dimensional-stability problem of uranium may be solved by (a) controlling the grain size; (b) alloying to form a stable isotropic phase; (c) alloying to decrease the high-temperature strength; (d) alloying to decrease the density.

11. Based on the data in Table 5·4, which chemical reactions are correct as written?

(a)	$2K + Ag_2O \rightarrow K_2O + 2Ag$
(b)	$2K + Na_2O \rightarrow K_2O + 2Na$
(c)	$Pb + ZnO \rightarrow PbO + Zn$
(d)	$Pb + Ag_2O \rightarrow PbO + 2Ag$

12. Thermal stress is the result of (a) high temperatures; (b) radiation damage; (c) thermal gradients in materials; (d) thermal expansion of materials.

6

NUCLEAR-REACTOR FUELS

6·1 Introduction

A nuclear reactor is a system usually consisting of a moderator and a fuel containing fissionable material, together with coolant and structural members, in which a self-sustaining chain reaction can be maintained. What are the physical properties of these reactor components? How does one fabricate uranium or thorium? What fuel materials can we use at high temperatures? So far in this text, we have examined the physics of the nuclear processes, the functions of each of the reactor components, and the materials problems involving these components. In this chapter and the two that follow, we shall examine the properties of important reactor materials, such as uranium, thorium, aluminum, zirconium, stainless steel, beryllium, graphite, etc. From the properties of materials, one can learn and understand their uses and limitations and can thereby be in a position to use them to design a complex mechanism.

The heart of a nuclear reactor is the region containing the fuel. Nuclear fuels must be composed, at least in part, of fissionable material. Therefore, a fuel must contain uranium 235 (U^{235}), plutonium, or uranium 233 (U^{233}), the three most readily available fissionable materials. U^{235} is the only fissionable material which is commonly found in nature. Natural uranium contains approximately one atom of U^{235} for every 140 parts of U^{238}. U^{233} is prepared artificially by the neutron irradiation of thorium, whereas plutonium is prepared by irradiating U^{238} with neutrons.

Before the student starts reading the following sections, it is recommended that he review the discussion of the functions of fuel elements as presented in Chap. 2.

PART I. URANIUM

6·2 Useful forms

Natural uranium, containing about 0.7 per cent of the fissionable isotope U^{235}, was used as a fuel in the form of massive metal slugs in the first nuclear reactors. Current reactors designed to operate at high flux levels or high temperatures use uranium enriched in U^{235} and diluted by alloying with some low-cross-section metal such as aluminum or zirconium. In other reactors, uranium may be incorporated in low-melting-metal eutectics or as a part of a molten salt. It may also be used as an oxide or carbide embedded in metal, graphite, or suitable ceramic material.

6·3 Abundance and availability

Uranium is very widely distributed in nature. It is easy to detect and estimate even in minute quantities, because uranium and its disintegration products, which are always associated with it in nature, are radioactive. The mineral containing the richest amount of uranium is pitchblende; however, the principal uranium-bearing deposits in the United States are made up of uranium carnotite and roscoelite. The uranium content of these last two ores is generally low, averaging about 0.2 per cent U_3O_8.

The two richest deposits of uranium are those at Great Bear Lake in Canada and at Katanga in the Belgian Congo. Other important ore deposits are in Czechoslovakia, the United States, and South Africa. The flow sheet for production of U_3O_8 from Canadian pitchblende has been presented previously, in Chap. 4.

6·4 Physical properties

Many of the important physical properties of uranium are given in Table 6·1. The physical properties reported by different experimenters will vary because these properties are affected by the purity of the metal. The values reported are selected as the best available at the present time.

6·5 Crystallography

Three different crystal forms of uranium metal are known, namely, alpha (α), beta (β), and gamma (γ). The alpha phase, which is orthorhombic (the three crystal axes are perpendicular to one another, and the lattice parameters are different in the three perpendicular directions), is stable up to 660°C. The unit-cell dimensions are $a_0 = 2.854$ Å, $b_0 = 5.867$ Å, and $c_0 = 4.957$ Å. The beta phase exists between 660 and 760°C. Its crystal structure is tetragonal (of three perpendicular directions, two have the same lattice parameter); $a_0 = b_0 = 10.590$ Å, and $c_0 = 5.634$ Å. Each unit cell contains 30 atoms.

Table 6·1 Physical properties of uranium

Density, gm/cm^3 ....19.13
Melting point...................2071 ± 2°F, 1133 ± 1°C
Boiling point ....7050°F, 3900°C
Electrical resistivity at
   25°C, microhm-cm ....2–4 × 10^4
Thermal conductivity, cal/cm-sec-°C:
    75°C 0.062
    400°C 0.078
Thermal expansion per °C, 25–125 × 10^{-6}
  Direction parallel to axis:
    100.......................21.17
    010...................... −1.5
    001.......................23.2
    Volume 45.8
Allotropic transformations:
  Alpha to beta ....1225°F, 663°C
  Beta to gamma............1407°F, 764°C

From 760°C to the melting point, the gamma phase is stable. It has a bcc structure, with $a_0 = 3.474$ Å.

6·6 Mechanical properties

Since uranium is a highly anisotropic material, the mechanical properties are affected considerably by orientation, as a result of fabrication and heat-treatment. For example, Table 6·2 shows the tensile properties of uranium

Table 6·2 Tensile properties of uranium extruded at various temperatures

Extrusion temperature, °F	Ratio	Tensile strength, 10^3 psi	Elongation, per cent
390	2:1	127	6
750	2:1	125	11
750	8:1	103	18
930*	8:1	96	13
1110	14:1	96	20

* Sample was high in impurities.

after extrusion at various temperatures and different reduction ratios (in extrusion, the reduction ratio is defined as the cross-sectional area before reduction over the final area). The results show that the strength properties decrease and the ductility increases as the temperature of the alpha extrusion is raised.

In addition to its dependence on orientation, uranium is very sensitive to impurities such as carbon and hydrogen. The tensile strength of as-cast rods of uranium will increase by 50 per cent when the carbon content is raised from 60 to 550 ppm. The sensitivity of uranium to hydrogen impurities is much more pronounced. Results indicate that as little as 2 ppm of hydrogen will markedly reduce the ductility of uranium. However, the deleterious effects of hydrogen can be removed by vacuum annealing; the better the vacuum, the greater the improvement in ductility.

6·7 Melting and casting

Melting and casting of uranium are made extremely difficult by the high degree of chemical reactivity of uranium with the atmosphere and with most crucible materials. The first problem can be solved by melting in vacuum, in a purified and dried inert atmosphere, or under a suitable slag or flux. The second problem has been partially solved with some sacrifice of metal purity. The usual crucible materials for uranium are graphite, beryllia, and thoria.

6·8 Forming and fabrication

Early efforts at fabrication by the Manhattan Project and later by the USAEC were concentrated on extrusion and rolling, and both methods were developed to a high state of efficiency. More recently, other methods of fabrication including forging, pressing, swaging, and drawing have been investigated and workable techniques established.

The ease of fabrication of uranium depends greatly on the temperature. In the gamma phase, uranium is so soft and plastic that rolling and swaging are difficult. The beta phase is generally harder and more brittle than the other two phases, but, with close temperature control, some rolling and extrusion have been accomplished in the beta regions. Most of the fabrication, however, is done in the alpha phase, where the metal is generally easy to fabricate. During most fabrication processes, care must be exercised to prevent oxidation. For this reason, uranium is rarely fabricated "bare" but is usually protected by a cladding of ductile metal such as copper or steel.

6·9 Joining or welding of uranium

While the joining of uranium has received considerable attention at various times, there is little information on completely satisfactory techniques. Use of any of the conventional techniques is complicated by the reaction of uranium with the atmosphere. Uranium has been welded by the *Heliarc process* (helium-arc welding) and the shielded-arc consumable electrode process (argon atmosphere). The electrode in the latter case was uranium wire.

Joining uranium to uranium by brazing or soldering is handicapped by the formation of brittle compounds between uranium and the joining metals.

Some acceptable joints have been made by first plating the uranium with silver or nickel and then joining the silver or nickel surfaces together by conventional methods.

6·10   Aqueous corrosion of uranium

Uranium reacts with steam between 150 to 250°C to form the stable oxide UO_2 and the stable hydride UH_3, according to the reaction

$$7U + 6H_2O_{gas} \rightarrow 3UO_2 + 4UH_3 \qquad (6·1)$$

At 600 to 700°C (1112 to 1292°F) the reaction products are reported to be pure UO_2 and hydrogen. Results indicate that the reaction with steam is much more vigorous than with oxygen under the same temperature conditions. Furthermore, the reaction rate between uranium and hydrogen is at all temperatures considerably higher than that between uranium and water.

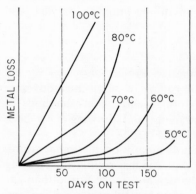

Two types of corrosion of uranium occur in water, depending on whether a protective film of UO_2 is formed and retained on the metal surface. In air-saturated distilled water, protective films are formed during the early stages of corrosion at 50 to 70°C, as shown in Fig. 6·1. During the later stages of corrosion (after 50 to 100 days), the reaction rate increases considerably. At the time when the reaction rates increase, it has been observed that the thin protective oxide films break down to form areas where the corrosion product is porous and black. These areas then spread, eventually enveloping the entire sample. The theory of this type of corrosion is that ionic hydrogen migrates through the thin oxide films which were formed during the early stages and produces uranium hydride between the metal and the oxide. Such a hydride layer spoils the adherence of the oxide to the metal, and its protectiveness disappears.

Fig. 6·1   Schematic representation of the corrosion of uranium in aerated distilled water. Note: Protective films are responsible for the low corrosion rate during the early stages at 50, 60, and 70°C.

The same mechanism is believed to describe the formation of uranium hydride in defectively jacketed pieces of uranium. When the jacket is not bonded to the uranium piece, water enters through the defect and reacts with the uranium to liberate hydrogen; the hydrogen is believed to diffuse along the surface of contact between jacket and uranium and may start reacting with the uranium at a considerable distance away from the jacket perforation. This distance is determined by such things as the temperature and the "induction period," which has been observed prior to the initial reaction between hydrogen and uranium. If the reactions are allowed to proceed in unbonded jacketed pieces in water, eventually water reaches the site where uranium hydride exists and converts it to uranium oxide. Thus, if such pieces are examined when a local swelling is first observed, the corrosion product is UH_3; after longer exposure, the larger swellings are composed of UO_2.

Alloying can be effective in stabilizing the protective oxide film on uranium metal surfaces. A uranium 5 weight per cent–zirconium alloy, for example, requires a very much smaller concentration of oxygen in the water to provide the highly corrosion-resistant surface film. Thus, in actively boiling distilled water in contact with air, this alloy develops the typical thin film showing interference colors (light reflection) and a very low corrosion rate. Eventually, these films seem to break down in much the same way as does the film formed on unalloyed uranium in air-saturated water at lower temperatures. However, when *oxygen-free* gas is bubbled through the boiling distilled water, the corrosion rate is nearly the same as that of unalloyed uranium, and the thin, protective film is not observed.

There are three general classes of alloys which appear to be able to form and to retain the protective oxide films, even in oxygen-free water, at temperatures as high as 350°C:

1. Certain metastable gamma-phase alloys
2. Certain supersaturated alpha-phase alloys
3. Certain intermetallic compounds

Typical of alloys of the first class are those containing more than about 7 weight per cent molybdenum or niobium. The supersaturated alpha-phase alloys characteristically contain small amounts of niobium. Typical alloys of this series are:

1. U—3 weight per cent Nb
2. U—1.5 weight per cent Nb–5 weight per cent Zr
3. U—3 weight per cent Nb–0.7 weight per cent Sn

The third type of corrosion-resistant uranium alloys is typified by the stable epsilon phase U_3Si in the U-Si system.

6·11   Uranium compounds

Uranium compounds are generally considered for use in elevated-temperature heterogeneous reactors. The desirable characteristics of a fuel compound are as follows:

1. High melting point and an absence of phase transformations.
2. Low neutron absorption cross section for the nonfissionable constituent.

3. Chemical and metallurgical inertness with respect to the reactor coolant and fuel-element cladding.

4. The structure of the compound should be relatively insensitive to uranium depletion.

5. The compound should retain fission products and should be relatively stable with respect to their accumulation.

6. Mechanical strength, resistance to thermal shock, high thermal conductivity, and high density of uranium atoms are important physical properties.

The compounds UO_2 and UC fulfill many of these requirements. UO_2, and its solid solutions with ThO_2 and PuO_2, appears to be one of the best high-temperature fuels available despite its lack of structural strength, low uranium atom density, and poor thermal conductivity. The relatively high thermal conductivity of uranium carbide has attracted attention, and a considerable amount of data concerning this compound should be available shortly.

The effects of irradiation on bulk UO_2 have been extensively studied by the Westinghouse Bettis Laboratory. The purpose of the study was to evaluate the reactor performance on bulk UO_2 as a fuel in a pressurized water-type reactor. From this study it was concluded that bulk UO_2 sheathed in metal tubes is suitable as a fuel material in high-temperature pressurized water reactors up to relatively moderate burnups.

Uranium compounds are also considered for use in dispersion-type core fuel elements. A dispersion type of fuel alloy is a two-phase alloy with particles of the fuel-bearing phase dispersed in a well-behaved metal matrix. To minimize radiation damage and provide metallic properties, the nonfissionable metal must predominate in volume and exist as a continuous matrix surrounding the fissionable phase. In effect, the matrix metal acts as a structural material in the fuel element. Typical dispersion-type fuel systems considered include UO_2 dispersed in stainless steel, Nichrome, aluminum, Zircaloy, or graphite and uranium carbide dispersed in stainless steel, zirconium, or Zircaloy. The record of successful performance of dispersion-type fuel elements confirms the theoretical advantages predicted in the development of the dispersion-fuel concept.

PART 2. PLUTONIUM

6·12   Unique properties

Plutonium is not found to any appreciable extent in nature but is mainly formed by the absorption of neutrons in U^{238}. The metal was first discovered in 1940, approximately 5 years before the first atomic bomb containing plutonium was exploded. At present, plutonium is being used as the fuel for the Los Alamos fast reactor. Its uniquely high average yield of approximately three neutrons per Pu^{239} fission makes it particularly applicable to a breeding operation. The full benefit of the high neutron yield can be effectively realized, but only if the bombarding neutrons that cause fission are "fast," i.e., have high energy. Because the characteristics of plutonium are exceptionally promising for power-breeder requirements, the importance of this element, both as fuel and as product, may soon transcend that of all other fissionable materials used in reactors.

6·13 Extraction and purification

Because plutonium is a synthetic element, the initial steps involved in winning it from its source materials are radically different from the processing schemes discussed in Chap. 4, such as roasting and leaching in the extraction of naturally occurring metals. The reduction to metal and purification of plutonium are, however, similar to the methods used for the ultimate reduction and purification of a number of other metals. The details of these techniques for plutonium production have not appeared generally in the unclassified literature. One further point: extreme care must be taken in handling plutonium because of the high toxicity of the material.

6·14 Physical properties

Plutonium is undoubtedly the most complicated, interesting, and exasperating metal yet known to man and produced in any quantity sufficient for metallurgical investigations. It has six allotropic modifications. The first

Table 6·3 Phase-transformation data on plutonium metal

Transformation	Temperature, °C	Volume change, %
α–β	122	8.9
β–γ	205	2.4
γ–δ	318	6.8
δ–δ'	452	−0.1
δ'–ϵ	479	−3.0
ϵ–liquid	640	

transformation occurs at 122°C, and the metal melts at about 640°C; thus, in a span of about 500°C, five allotropic transformations occur. The phase-transformation data are summarized in Table 6·3. It should be noted that the δ' phase exists over a temperature range of only some 30°C. This explains why it was missed in earlier investigations. It should also be noted that, in two of the transformations, the volume change is negative.

The crystal's structures and several physical properties of the six phases are summarized in Table 6·4. The alpha- and beta-phase structures are extremely complicated, and neither has been worked out. It is known,

Table 6·4 Crystal structure and physical properties of high-purity plutonium metal

Phase	Crystal structure	Lattice parameters, Å	Density at 25°C, gm/cm³	Av. linear expansion coefficient, °C	Electrical resistivity microhm-cm
α	(Unlike uranium)	?	19.8	55×10^{-6}	150 at 25°C
β	Complex unknown	?	17.8	35×10^{-6}	110 at 200°C
γ	Face-centered orthorhombic	$a_0 = 3.1587$ $b_0 = 5.7682$ $c_0 = 10.162$	17.14	36×10^{-6}	110 at 300°C
δ	Face-centered cubic	$a_0 = 4.6371$	15.92	-21×10^{-6}	102 at 400°C
δ'	Face-centered tetragonal	$a_0 = 4.719$ $c_0 = 4.453$	16.01		
ϵ	Body-centered cubic	$a_0 = 3.639$	16.48	4×10^{-6}	120 at 500°C

however, that alpha-phase plutonium does not have the same structure as alpha uranium.

6·15 Alloying characteristics

Plutonium is a relatively good alloy former. It is completely soluble with almost all metals in the liquid state, but few elements are soluble to any extent in any of the solid plutonium phases. Plutonium is an active compound former. At the present time, there are of the order of 50 intermetallic compounds known with this element.

PART 3. THORIUM

6·16 Thorium as a fuel element

Because of its low strength, poor resistance to corrosion, and high cost, thorium has not received consideration as a structural material; hence, until recently, little incentive has existed for producing it in quantity. Some possibility of using thorium as a minor constituent in certain alloys has developed, but the principal reason for the current active interest in thorium is its nuclear properties. The nucleus of the thorium atom Th[232] can undergo a reaction with a neutron to form Th[233], which is relatively short-lived. Subsequent radioactive disintegrations of this isotope produce relatively long-lived U^{233}, which can be caused to fission and, accordingly, may serve as an atomic fuel in a manner similar to U^{235} and Pu^{239}.

6·17 Occurrence and purification

Thorium-containing ores are distributed all over the earth; the concentration of most of the deposits, however, is fairly low. Monazite sand, as found in Brazil and India, while consisting mainly of rare-earth oxides, contains significant amounts of thorium in the form of the oxide ThO_2, along with fairly small amounts of the uranium oxide U_3O_8. Thorium can be extracted by treating the ore with hot sulfuric acid, which dissolves the thorium rare earths. Separating thorium from cerium, lanthanum, and yttrium earths presents many difficult technical problems, none of which will be discussed here. In spite of the complicated nature of the extractive procedures, the thorium products are very pure.

6·18 Physical properties

Pure thorium metal resembles steel in appearance, but its hardness is low, in the range of that of silver. Table 6·5 lists some of the physical properties

Table 6·5 Physical properties of thorium

Density, gm/cm^3 . . . . .11.71
Melting point. . . . . .3075° $\pm$ 20°F, 1690 $\pm$ 10°C
Boiling point . . . . .3000°C
Coefficient of thermal expansion, per °C $\times$ 10^6:
    30–100°C 11.5
    30–1000°C 12.5
Electrical resistivity, at 20°C,
    microhm-cm. 18
Crystallography:
    Face-centered cubic . . . . . .a_0 = 5.087 Å
    Above 1400 $\pm$ 25°C
    Body-centered cubic . . . . . .a_0 = 4.11 Å
Thermal conductivity, cal/sec-cm-°C:
    100°C 0.090
    650°C 0.108

of thorium. At room temperature, thorium has an fcc unit cell. Recent experiments indicate a transformation to a bcc phase at 1400 $\pm$ 25°C.

6·19 Mechanical properties

The mechanical properties of thorium are strongly influenced by impurities. In particular, carbon, even in small amounts, changes these properties extensively. The effects of carbon, oxygen, and nitrogen additions on the tensile strength and ductility of arc-melted iodide metal are indicated in

Fig. 6·2. Oxygen and nitrogen have little effect on the strength of the metal, but carbon increases the strength markedly. As shown in Fig. 6·2, 0.2 per cent carbon increases the strength from 20,000 to 55,000 psi and decreases the possible reduction in area from approximately 60 to 30 per cent.

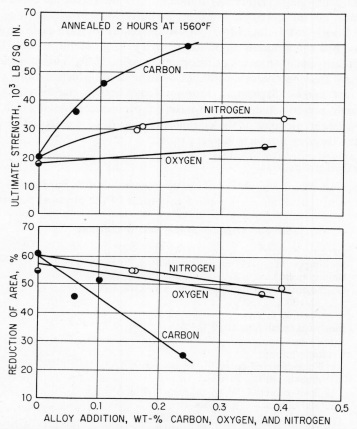

Fig. 6·2 The effects of carbon, oxygen, and nitrogen on the tensile strength and ductility of iodide thorium.

Thorium work hardens rapidly when subjected to cold working. The increase in hardness is small, however, and thorium can therefore be cold-worked to quite high reductions without intermediate anneals. For example, thorium can be rolled without intermediate annealing steps to 90 per cent reduction in area. The annealing treatment for work-hardened thorium generally consists of a 1-hr anneal at 1450°F in an inert atmosphere or in vacuo.

6·20 Fabrication

Because of its high melting point and reactivity, thorium is a difficult metal to melt. It is necessary to melt it in a vacuum or inert atmosphere to avoid contamination with oxygen and nitrogen. The metal is usually heated by induction, but small melts have been made in arc furnaces.

The high ductility of thorium permits standard working procedures at room temperature. The high degree of oxidation at elevated temperatures makes protective precautions, such as providing for an inert atmosphere or jacketing, necessary for any hot-working procedure. There is no brittle temperature range above room temperature, although castings have a tendency to crack when cold-worked. However, once the cast structure has been broken up, the metal can be fabricated at room temperature.

The techniques for welding and brazing thorium are not yet well developed, and brittleness and cracking frequently occur. Alloying with 2.5 atomic per cent of molybdenum, tungsten, or niobium improves the welding characteristics of the metal.

6·21 Corrosion behavior

The corrosion resistance of thorium in air and boiling water is low. Oxidation in air, even at low temperatures, is fairly rapid. Thorium must be protected by a cladding material if it is to be used in a reactor with boiling water as the coolant. Alloys containing 25 to 30 weight per cent zirconium are, however, corrosion-resistant in boiling water. Thorium shows excellent corrosion resistance in a eutectic sodium-potassium alloy and also in lithium at temperatures as high as 1110°F.

6·22 Alloying characteristics

With the exception of carbon and indium, small additions of up to 5 atomic per cent of most elements are ineffective in strengthening bomb-reduced thorium. In fact, small additions of titanium, zirconium, and niobium decrease the as-cast hardness of bomb-reduced thorium markedly, probably through a scavenging action on the carbon and oxygen in the bomb-reduced material. Thorium forms intermetallic compounds with most of the metals. Equilibrium diagrams for most of the binary alloys are described in the literature.

PROBLEMS

As a means of summarizing the chapter, the student is asked to make a large table comparing the properties of the nuclear-fuel metals U, Pu, and Th. The list of properties to be compared should include (a) nuclear cross-section data; (b) metallurgical processing methods; (c) physical properties; (d) mechanical properties; (e) fabrication techniques; (f) corrosion behavior; (g) alloying characteristics.

7

CONSTRUCTION MATERIALS

7·1 Introduction

The materials used in the construction of a reactor are chosen to perform certain functions. The particular properties of certain materials make them especially desirable for use as fuel elements, moderators, or core materials. Similarly, certain properties of aluminum, stainless steels, and zirconium make these metals and alloys particularly useful as constructional materials. In this chapter we shall consider the properties of these materials with special reference to the specialized functions which they must perform.

The structural components of a reactor include the ductwork, piping, control mechanisms, valves, baffles, control rod sleeves, and many more. The choice of material for each of these components must be considered from the viewpoint of the following properties:

1. Mechanical strength and creep strength. The reactor and its components constitute a mechanical structure which must support itself under the elevated temperature and stress conditions of reactor operations.
2. Corrosion resistance. Components must resist corrosion attack by coolants and atmospheres.
3. Heat conduction. Efficient transfer of heat from the reactor is a major requirement.
4. Neutron cross section.
5. Radiation damage. The material must resist radiation-damage effects, such as decrease in ductility and change in dimensions.
6. Purity. Small amounts of impurities may impair ductility or reduce neutron economy.

PART I. ALUMINUM AND ITS ALLOYS

7·2 Uses

Aluminum has been used in a number of relatively low-temperature water-cooled nuclear power reactors. It is used for simple structural members and reactor assemblies, such as fuel elements, coolant passages, and control elements, in order to impart additional strength to these assemblies and consequently strengthen the structure as a whole.

7·3 Fabricability

One of the advantages of using aluminum as a constructional material in reactors is the ease with which it may be produced and fabricated. Pure and mildly alloyed aluminum may be formed into almost any shape, and the harder alloys can often be deformed severely if preheated before deformation. Riveting, welding, forging, and machining are relatively easy, and cast shapes can be produced if proper care is taken to prevent porosity. This porosity arises from the fact that hydrogen is increasingly soluble in the melt as the temperature is raised but is, for all practical purposes, insoluble in the solid metal. The porosity which results from the precipitation of hydrogen during solidification may be prevented by removing the hydrogen by bubbling nitrogen through the melt just before pouring.

7·4 Physical properties

The physical properties of aluminum are given in Table 7·1. As seen in the table, the thermal conductivity of aluminum is particularly good. In addition, the thermal neutron absorption cross section of aluminum is low, and in this respect it is particularly desirable for thermal reactors.

Table 7·1 Physical properties of aluminum

Thermal neutron cross section, barns/atom 0.215
Density, gm/cm^3 . . . . .2.70
Melting point. . . . . .1220°F, 660.2°C
Boiling point . . . . .4225°F, 2327°C
Coefficient of linear thermal expansion per °C:
    20–100°C . . . . .23.8 × 10^{-6}
    20–400°C . . . . .26.7 × 10^{-6}
Electrical resistivity, microhm-cm:
    20°C 2.66
    100°C 3.86
    400°C 8.0
Thermal conductivity, cal/sec-cm-°C:
    20°C 0.503
    100°C 0.503
    400°C 0.546
Lattice structure . . . . .Face-centered cubic
Lattice constant. . . . .$a = 4.0489$ Å

7·5   Radiation-damage effects

Since the neutron irradiation of metals results in the formation of vacancies and interstitials, it would be expected that a metal would show little or no observable effects if irradiated in a temperature range in which vacancies and interstitials would be annihilated by annealing. High-purity aluminum is known to anneal itself at room temperature; as expected, no radiation damage is noted when the metal is irradiated at this temperature. It must be remembered that the addition of small amounts of impurities can increase the annealing temperature and, thereby, increase the effects of irradiation at room temperature. The foregoing may be compared with the effect of irradiation on copper, which does not anneal at room temperature. In this case, the hardening which occurs upon irradiation is comparable with the maximum hardness change shown by cold work.

7·6   Mechanical properties

Commercially pure aluminum (99.1 per cent pure) has a tensile strength of 13,000 psi, with 30 to 35 per cent elongation. Cold work may increase the

Table 7·2   Chemical compositions of some wrought aluminum alloys

Alloy	Silicon	Copper	Man-ganese	Mag-nesium	Chro-mium	Iron	Zinc	Titanium
2S	*	0.20	0.05	....	....	*	0.10	
3S	0.60	0.20	1.5	....	....	0.70	−10	
17S	0.80	4.5	1.0	0.80	0.10	1.0	0.10	
24S	0.50	4.9	0.90	1.8	0.10	0.50	0.10	
52S	†	0.10	0.10	2.8	0.35	†	0.10	
61S	0.80	0.40	0.15	1.2	0.35	0.70	−20	0.15
63S	0.60	0.10	−10	0.85	0.10	0.35	0.10	0.10

* 1.0Si ± Fe.
† 0.45Si ± Fe.

tensile strength to about 24,000 psi, with a reduction of elongation to about 5 per cent. The mechanical strength of pure aluminum can be improved substantially by alloying. The composition of typical alloys is listed in Table 7·2. Detailed information on the strength of alloys at room temperature and at 400 and 600°F is presented in Fig. 7·1. The 2S, 3S, and 52S alloys are strengthened by solid-solution hardening and are not heat-treatable. Their strength may be increased, however, by cold work. The heat-treatable alloys, such as 17S, 24S, 61S, and 63S, are of the precipitation-hardening type. The classic aluminum alloy with copper is hardened by quenching from an elevated temperature in order to retain the copper in solution, followed by "aging" at some intermediate temperature to precipitate $CuAl_3$ and thus harden the alloy. The addition of copper, it should be noted, reduces the corrosion resistance of aluminum and increases its neutron cross section.

The fact that precipitation-hardening alloys are aged at 250 to 400°F indicates that such alloys cannot be used at these temperatures because of the softening which results from overaging. This brings us to an important point: the effect of temperature on mechanical strength and creep strength

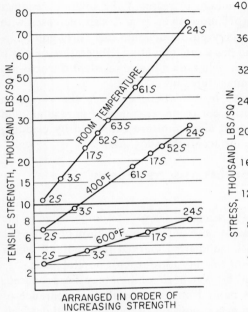

Fig. 7·1   Relative tensile strength of wrought aluminum alloys at room temperature, 400, and 600°F.

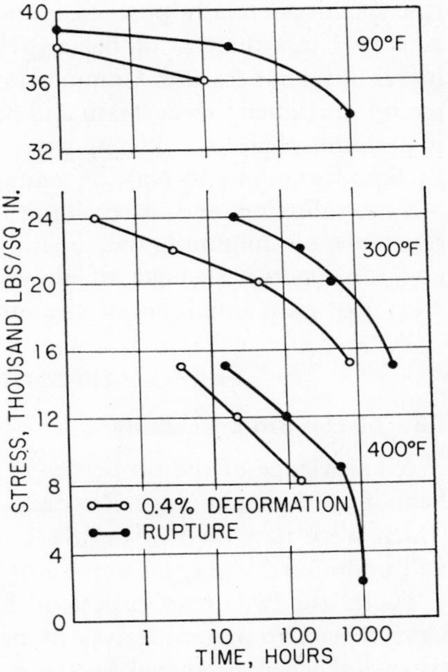

Fig. 7·2   Design curves for 63S aluminum alloys at 90, 300, and 400°F.

of aluminum and its alloys. More than any other factor, the loss of strength by aluminum at elevated temperatures restricts its use in nuclear reactors. The melting temperature of pure aluminum is 1220°F, and the addition of alloying elements reduces the melting temperature to as low as 950°F in the 17S alloy. These are, of course, the extreme upper limits for structural strength. However, most aluminum alloys suffer severe reduction in tensile and creep strengths as low as 400°F. Figure 7·2 illustrates the effect of temperature in limiting the design stress on the 63S alloy. It will be noted that a ,stress of only 10,000 psi at 400°F will cause a deformation of 0.4 per cent in 30 hr and will cause complete failure (rupture) in only 200 hr. It can be appreciated, therefore, that other materials must be considered for load-carrying members which must operate at temperatures above 300 to 400°F.

7·7   Corrosion resistance

Commercially pure aluminum has very good corrosion resistance in atmospheric or marine environments. Its corrosion resistance in water is good in the cold-worked, as well as the annealed, condition. The corrosion rate of commercially pure 2S aluminum increases if the solution becomes acid (pH less than 5) or basic (pH more than 7) and also increases with increasing temperature. Commercially pure aluminum has fairly good corrosion resistance to wet steam and has a safe operating temperature of 800°F in an atmosphere of oxidizing gases. Aluminum is also resistant to corrosion by liquid sodium and NaK at temperatures up to 200°C.

Since alloying and precipitation heat-treatments reduce the corrosion resistance of aluminum, the superior corrosion resistance of pure aluminum and the superior strength of alloyed aluminum are combined by cladding a layer of pure aluminum on the surface of the alloy grade.

PART 2.   ZIRCONIUM AND ITS ALLOYS

7·8   Zirconium processing

A knowledge of the properties of zirconium is of vital importance in the field of reactor technology. Zirconium has a unique combination of properties which were first used in the STR (submarine thermal reactor) and which will undoubtedly play an important role in future reactor development.

There are two main processes by which zirconium may be produced. Crystal bar zirconium derives its name from the crystalline appearance of the metal when produced by the decomposition of zirconium iodide. This product, also called "iodide zirconium," is very pure and quite expensive. The majority of zirconium used in reactors today is "sponge zirconium," which is produced by the famous Kroll process. In this process, also used for titanium, zirconium chloride is reduced to metallic zirconium by magnesium to produce a spongy mass—hence the name sponge. Zirconium may be induction-melted in graphite crucibles, but the increase in carbon content from such melting results in increased difficulty in fabrication and in decreased corrosion resistance. It can be arc-melted in water-cooled copper crucibles by using either tungsten electrodes or the zirconium itself as an electrode in the "consumable electrode" process.

7·9   Physical properties

Table 7·3 lists some of the general physical properties of zirconium. Of particular interest is the thermal neutron cross section for zirconium, which is only 0.18 barn per atom. This low value is one of the main reasons for the popularity of this metal as a constructional material and cladding for fuel elements. Since zirconium is generally alloyed for use in reactors, the effects of the alloying elements on the cross section must be considered. The addition of nickel, chromium, and iron for improved corrosion resistance results in an increase in the thermal neutron cross section. Tin can be used liberally as an alloying element, since it has a low cross section.

Table 7·3 Physical properties of zirconium

Thermal neutron absorption cross section,
    barns/atom 0.18
Density, gm/cm³ 6.50
Melting point . . . . .3355° ± 75°F, 1845° ± 25°C
Coefficient of linear thermal expansion,
    per °C at 250°C:
      C axis . . . .6.15×10^{-6}
      a axis . . . .5.69×10^{-6}
Electrical resistivity, microhm-cm:
      20°C 42
      100°C 59
      400°C 100
Thermal conductivity, cal/sec-cm-°C
    at room temperature 0.05
Lattice structure (up to 863°C) . . . . .Hexagonal close-packed
Lattice constants at room temperature . . . .$a_0 = 3.2321$ Å
                                                          $c_0 = 5.1474$ Å

In any discussion of the cross section of zirconium, it should be remembered that zirconium ores normally occur as a mixture of zirconium and hafnium oxides. In chemical behavior hafnium and zirconium are almost exactly alike, and consequently elimination of hafnium from zirconium is very difficult. This separation is critical, because hafnium has a high cross section.

7·10 Radiation damage

Unlike aluminum, zirconium is significantly affected by neutron irradiation. Annealed zirconium increases 100 Bhn (Brinell hardness number) after a 10^{19} nvt exposure. Such an exposure also increases yield and tensile strengths about 10,000 psi. Cold-worked zirconium increases only slightly in hardness during irradiation, as would be expected, since vacancies and interstitials would not be expected to increase the hardness of a cold-worked metal.

7·11 Mechanical properties

Typical values of the mechanical properties of zirconium are found in Table 7·4. Cold work results in a pronounced increase in the tensile strength of zirconium and its alloys. Strength increases of 50,000 to 100,000 psi are

Table 7·4 Mechanical properties of zirconium

Effect of Processing and Melting Practice

Type of Zr and melting method	Fabrication history	Strength, psi (0.2% offset)		Per cent elonga-tion	Hardness Rockwell A
		Yield 37,600	Tensile 61,500		
Sponge, induction-melted in graphite crucibles	Rolled at 1560°F, annealed 30 min, air-cooled			21	54
Sponge, induction-melted in graphite crucibles	Forged and rolled at 1830°F, cold-rolled 30%, annealed 1 hr at 1300°F	35,800	56,400	24	50
Sponge, arc-melted	Forged and rolled at 1830°F, cold-rolled 30%, annealed 1 hr at 1300°F	37,600	63,300	30	54
Iodide, arc-melted (0.004 weight % nitrogen)	Forged and rolled at 1450°F, cold-rolled 30%, annealed 1 hr at 1300°F	16,400	35,500	36	34
Iodide, arc-melted	Hot-rolled at 1400°F, cold-rolled 10%, annealed 20 hr at 1380°F	7,700	24,800	40	20
Iodide, arc-melted (0.002 weight % nitrogen)	Forged and rolled at 1450°F, cold-rolled 30%, annealed 1 hr at 1300°F	13,600	29,000	29	20

not uncommon. The effects of cold work are removed by annealing at 1200 to 1300°F.

7·12 Effects of impurities on mechanical properties

Pure zirconium is a soft, ductile metal and is easily fabricated. The addition of small amounts of iron, chromium, nickel, and tin increases the strength and reduces ductility correspondingly. With care, such alloys do not offer insurmountable difficulties in fabrication, but the presence of oxygen from the reduction or melting operation can seriously impair the ductility of zirconium and its alloys. Proper vacuum techniques and care in handling the zirconium chloride before reduction are necessary to keep oxygen contamination to a minimum. The effects of oxygen on the mechanical properties of zirconium are shown in Table 7·5. The amount of oxygen in

Table 7·5   Effect of oxygen on hydrogen-free zirconium

Atomic % oxygen	Ultimate strength, psi	0.2% yield psi	Elongation, % in./in.	Hardness Rockwell A
None	29,000	7,000	32	20
0.5	39,000	26,000	16	36
1.0	50,000	34,000	7	46
1.5	59,000	42,000	5	52
2.0	70,000	49,000	4	57
2.5	81,000	56,000	3	60

the sponge can be roughly estimated from the hardness because of the sensitivity of the latter to oxygen content. The hardness also permits an estimate to be made of the fabricability of the metal. The effects of nitrogen on zirconium are similar to those of oxygen.

Small amounts of hydrogen in zirconium have a greater embrittling effect than oxygen or nitrogen. Hydrogen forms a brittle hydride which precipitates, during cooling from 600°F, in the form of thin plates. These plates cause brittleness. Brittleness is best described in terms of its effect on the impact strength, which is measured in terms of the number of foot-pounds of energy which a specimen of a given size absorbs when it is broken very rapidly. Hydrogen-free zirconium has an impact strength of 40 to 70 ft-lb. This impact strength is reduced to 10 ft-lb by the addition of only 60 ppm (0.006 per cent of hydrogen). Both pure zirconium and tin alloys of zirconium evidence this brittleness when slowly cooled from 600°F. Unlike oxygen or nitrogen, hydrogen is easily picked up by zirconium from furnace atmospheres, pickle baths, or descaling baths; but also, unlike nitrogen or oxygen, it can be removed by vacuum annealing at high temperatures.

7·13   Effects of temperature on mechanical properties

The strength of zirconium decreases rapidly with temperature, but it is greater than that of aluminum. The addition of oxygen increases the room-temperature strength substantially, although this effect is largely lost at elevated temperatures. The addition of tin as an alloy element, however, increases both room-temperature and elevated-temperature strength. The high-temperature short-time tensile strength of zirconium is also increased by molybdenum, titanium, and aluminum. However, reactor designs must be made on the basis of long-time tests, such as creep or stress-rupture tests. Such tests indicate that these alloys, including the tin alloys, have only about one-fifth the creep strength of stainless steels at a temperature of 900°F. These alloys are superior to aluminum in the intermediate-temperature range but do not have sufficient strength for operation at high temperatures. Largely for this reason, and also because the corrosion resistance of zirconium decreases markedly above 600°F, we must consider stainless steels, which will be discussed in Part 3.

7·14 Corrosion resistance

The corrosion resistance of zirconium is generally very good—in some cases phenomenal. This metal remains unattacked by many acids, chemical reagents, and sea water. It also has good resistance to liquid sodium and NaK up to 600°C.

Since the STR is water-cooled, the corrosion of zirconium in water and steam is of great importance. It will withstand pressurized water at 600°F for very long periods of time. The corrosion rate of zirconium is less than 10 mgm/decimeter² per month at 600°F, but it may corrode rapidly at 750°F. It is known that even small amounts of impurities can seriously impair its corrosion resistance in high-temperature water. Small amounts of nitrogen contamination are invariably associated with the sponge-production and melting operations. As little as 50 ppm of nitrogen observably increases the corrosion rate. The presence of carbon in zirconium results in the formation of zirconium carbide, which is apparently corroded rapidly if the carbon content exceeds 500 ppm (0.05 per cent).

Fortunately, alloying elements such as tantalum, niobium, and especially tin can be added to zirconium to offset the detrimental effects of nitrogen, carbon, and oxygen, Tin improves the corrosion resistance of high-carbon induction-melted zirconium, and the addition of 1 per cent tin for each 300 ppm of nitrogen results in an alloy of good corrosion resistance. However, the corrosion resistance of high-purity zirconium is not improved by tin additions.

It has been found that the rate of corrosion in water will sometimes increase rapidly after a period of exposure. This increased rate is accompanied by the formation of an adherent film of corrosion products, rather than the usual white powder associated with slower corrosion rates. The addition of iron, chromium, and nickel is known to prevent the failure which is associated with film formation. It should be remembered that most of the elements which might be found as impurities in zirconium do not have any effect on its corrosion resistance. These include oxygen, tungsten, vanadium, cobalt, molybdenum, lead, zinc, thorium, and beryllium.

PART 3. STAINLESS STEELS

7·15 Grades of stainless steel

As we have seen, each material which can be considered as a constructional material for reactors has a unique set of properties. The choice of the material to be used for a given component depends upon the properties of the material relative to the conditions under which it must operate. The advantages of stainless steels, as might be expected, arise from their excellent corrosion resistance and retention of strength at elevated temperatures. Since the thermal efficiency of reactors increases with the coolant temperature, it can be expected that the use of stainless steels will increase in the next few years.

Any discussion of stainless steels must be preceded by a description of the three main types with respect to the particular properties which make them useful. The three main types of stainless steel are the martensitic, ferritic, and austenitic grades.

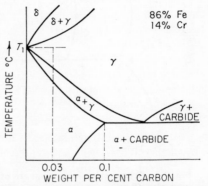

Fig. 7·3 Schematic phase diagram of iron plus 14 per cent chromium with carbon (martensitic stainless steels).

7·16 Martensitic stainless

The schematic phase diagram for the 14 per cent chromium martensitic stainless is shown in Fig. 7·3. This phase diagram is a pseudobinary. Actually, the diagram shows the phases present when carbon is added to alloys of iron and chromium in which the chromium content is held constant at 14 per cent. Because of the presence of the third element, pseudobinaries contain odd-shaped phase boundaries. For further discussion of this topic, the student should read a section on ternary diagrams in one of the basic metallurgy texts.

As shown in Fig. 7·3, a 0.03 per cent carbon alloy (with 14 per cent chromium, the balance iron) heated to T_1 assumes an fcc gamma structure. This structure in iron-base alloys is often called the austenite phase. On slow cooling, the structure transforms to the bcc phase. If the cooling is very rapid, a transition product is formed called *martensite*. This product is very hard in this particular alloy because carbon is present. The alloy can be hardened by quenching from the gamma phase and is, in fact, used for cutting edges.

7·17 Ferritic stainless

The second type of stainless contains about 18 per cent chromium and is called *ferritic stainless*. The phase diagram for this alloy is given in Fig. 7·4.

The 0.1 per cent carbon alloy (with 18 per cent chromium, the balance iron) does not transform to the gamma structure on heating and thus cannot be hardened by quenching, since there is no phase change. The alloy remains bcc within the heat-treating range. This bcc structure in iron-base alloys is called the alpha, or ferrite, phase—hence the name ferritic stainless. This grade of stainless can be hardened by cold work.

Both martensitic and ferritic stainless steels can be seen in a binary iron-chromium phase diagram in Fig. 7·5 without the complicating effects of

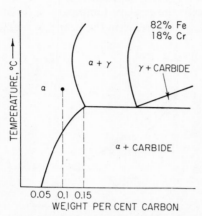

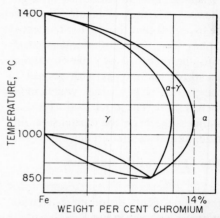

Fig. 7·4 Schematic phase diagram of iron plus 18 per cent chromium with carbon (ferritic stainless steels).

Fig. 7·5 Schematic iron-chromium phase diagram.

carbon. The martensitic steels, with less than 14 per cent chromium, transform from gamma to alpha on cooling, and the ferritic steels, with more than 14 per cent chromium, consist of ferrite up to the solidus.

7·18 Austenitic stainless

The *austenitic stainless* steels contain nickel in addition to iron and chromium, which makes the gamma (austenite) phase stable at room temperature. The best known of these steels is 18-8, which contains 18 per cent chromium and 8 per cent nickel. A schematic phase diagram is given in Fig. 7·6. Although the equilibrium phase diagram shows alpha stable at room temperature, the reaction is so sluggish that no heat-treatment can cause the transformation to alpha. For all practical purposes, the metastable austenite can be considered a stable structure. Since no transformations occur on cooling from the solution temperature, the alloy cannot be quench-hardened, but it can be strengthened by the addition of 0.1 per cent carbon and/or by cold work.

A disadvantage of austenitic stainless steels is that chromium carbides are precipitated at the grain boundaries upon heating to 800 to 1200°F. Since chromium is largely responsible for the corrosion resistance of stainless steels, the chromium which is precipitated in the form of carbides cannot participate in resisting corrosion. Consequently, the grain boundaries of such samples of 18-8 stainless steels are depleted in chromium and are subject to intergranular corrosion (grain-boundary corrosion). This susceptibility is frequently reduced by the addition of strong carbide-forming elements, such as niobium (columbium) or titanium. These elements tie up the carbon and thus reduce the tendency toward the formation of chromium carbides.

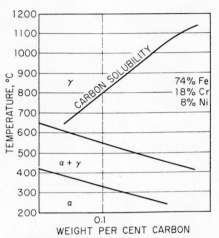

Fig. 7·6 Schematic phase diagram of iron plus 18 per cent chromium and 8 per cent nickel with carbon (austenitic stainless steels).

7·19 Composition of stainless steels

There are, of course, many grades of stainless steels which fall into the general classifications of martensitic, ferritic, and austenitic. A detailed summary of the compositions of available wrought stainless steels may be found in Table 7·6. It will be noticed that the 301 grade may contain as little as 6 per cent nickel. Austenitic steels, which are low in nickel, may partially transform to a low-carbon martensite on cold working. Austenitic stainless alloys are not ferromagnetic. However, severe cold work can transform some of the austenite to martensite and can thus cause the alloy to become ferromagnetic.

7·20 Physical properties

The range of physical properties for all types of stainless steels is listed in Table 7·7. In general, the martensitic and ferritic grades have better thermal and electrical conducting properties and a slightly higher melting point. The thermal neutron absorption cross section of stainless steels varies with composition and can be calculated from a knowledge of the cross sections of elements which compose each alloy. Since the thermal neutron cross sections of stainless steels are fairly high, the use of stainless in thermal reactors must be considered carefully. These alloys are particularly suited for

Table 7·6 The composition of wrought stainless steels

Class and composition limits	AISI and ASTM designations	Composition %			
		Carbon (maximum)	Chromium	Nickel	Other
Austenitic.....................	301	0.08–0.20	16–18	6–8	
Nonhardenable	302	0.08–0.20	17–19	8–10	
Nonmagnetic	303	0.20	17–19	8–10	Low P and S
Carbon = 0.25 (max.)	304	0.08	18–20	8–10	P = 1.0 (max.)
Manganese = 2.0 (max.)	305	0.08	17–19	10–13	
Silicon = 1.0 (max.)	308	0.08	19–21	10–12	
Phosphorus = 0.4 (max.)....	309*	0.20	22–24	12–19	
Sulfur = 0.4 (max.)	310	0.25	24–26	19–22	
Chromium = 16–26	316	0.10	16–18	10–14	Mo = 2 to 3
Nickel = 6–22	317	0.10	17.5–20	10–14	Mo = 3 to 4
	318	0.08	17–19	13–5	Mo = 2 to 2.75 Nb = 10 × C
	321†	0.10	17–19	8–11	Ti = 5 × C
	330	0.10	14–16	33–36	
	347‡	0.10	17–19	8–12	Nb = 10 × C
Martensitic....................	403	0.15	11.5–13	....	Turbine quality
Hardenable................	410	0.15	11.5–13.5		
Magnetic..................	414	0.15	11.5–13.5	1.25–2.5	
Carbon = 0.15–1.20	416	0.15	12–14		
Manganese = 1.0 (max.)	418	0.15	12–14	0.5	W = 2.5 to 3.5
Silicon = 1.0 (max.)	420	0.15	12–14		
Phosphorus = 0.04 (max.)...	431	0.2	15–17	1.25–2.5	
Sulfur = (0.04 max.)	440A	0.6–0.75	16–18		
Chromium = 4–18	440B	0.75–0.95	16–18		
Nickel = 2.5 (max.)	440C	0.95–1.2	16–18		
	501	0.10	4–6		
	502	0.10	4–6		
Ferritic	405	0.08	11.5–13.5		Al = 0.1 to 0.3
Nonhardenable	406	0.15	12–14		Al = 3.5 to 4.5
Magnetic..................	430	0.12	14–18		
Carbon = 0.35 (max.)	442	0.35	18–23		
Same limits on P, S, Mn, Si as martensitic	443	0.20	18–23		Cu = 0.9 to 1.25
Chromium = 14–18 .... Nickel = none	446	0.35	23–27		

* Made in stabilized grade by addition of Nb = 10 × C.

† Stabilization with titanium is undesirable from the standpoint of increasing the nuclear section of the alloys.

‡ Made in a free-machining grade by addition of Se = 0.15 to 0.35 and P = 0.11 to 0.17.

use in intermediate and fast reactors, where the somewhat higher-cross-section materials can be tolerated with little effect on neutron economy.

7·21 Radiation damage

Stainless steels show considerable effects after fast neutron irradiation. Hardness increases of 100 Bhn (Brinell hardness number) have been observed in annealed stainless after a neutron irradiation of 10^{19} nvt. The same amount

Table 7·7 Physical properties of stainless steels

Range of Values for All Grades of Stainless Steels

Density, gm/cm³ 7.6–7.98
Melting range ....2500–2750°F, 1370–1510°C
Electrical resistivity,
   microhm-cm:
      20°C 45–79
      400°C 88–105
      800°C 115–121
Thermal conductivity,
   cal/cm-sec-°C:
      100°C 0.032–0.087
      500°C 0.040–0.080

of radiation will increase the hardness of hardened stainless about 35 Bhn. These results are typical of the irradiation effects on annealed and cold-worked materials.

7·22 Mechanical properties

The mechanical properties of stainless steels, particularly at high temperatures, are the main reasons for their increasing application in jet engines, turbines, and nuclear power reactors. A vast amount of information is available on rupture, creep, and high-temperature strength of stainless steels. In addition, tensile tests, as a function of temperature, are of general value in classifying materials. Austenitic stainless (type 302) has a tensile strength of about 85,000 psi at room temperature. The tensile strength at 1000°F is 65,000 psi, and at 1200°F it is 50,000 psi. These results are indeed indicative of a useful high-temperature material. The creep strength at 1000°F is about 18,000 psi for a creep rate of 1 per cent in 10,000 hr. Under the same conditions, type 304 has a strength of 32,000 psi. As one can easily see, these alloys are certainly useful engineering materials at elevated temperatures, particularly in comparison with aluminum or zirconium.

7·23 Fabrication

Stainless steels, in general, are readily fabricated into a variety of shapes, and they can also be cast. One consideration in welding the 18-8 grade is that precipitation of carbides occurs in the heat-affected zone adjacent to the weld. This problem is attacked either by the use of a niobium-stabilized grade (such as type 347) or by solution treating at 2050°F after welding to dissolve the carbides, followed by quenching to keep the carbides in solution. Welding of the 28 per cent chromium alloy (ferritic) results in grain growth. This grain growth does not decrease corrosion resistance, but it is undesirable from the standpoint of mechanical properties.

7·24 Corrosion resistance

The corrosion resistance of stainless steels to various liquid and gaseous media is well known. Indeed, their name "stainless" refers to their overall corrosion resistance. The corrosion resistance of stainless steels arises from a phenomenon known as *passivity*. The steels resist corrosion when in the passive condition, which is characterized by an extremely thin film of oxygen atoms, or oxide, on the surface. The exact mechanisms by which the steels attain passivity is not known. However, experimental evidence indicates that chromium is the basic element for attaining passivity in steels. In general, corrosion resistance increases with increasing chromium content. Since the passive state is connected with an oxide film, strongly reducing conditions (lack of oxidizing conditions) increase susceptibility to attack. On the other hand, the presence of strongly oxidizing conditions causes extraordinary resistance to attack.

Chloride ions in salt solutions are detrimental to the corrosion resistance of stainless steels. This effect is generally small if the surface of the solution is exposed to air, because passivity can then be maintained. However, stagnant enclosed salt solutions do not contain much dissolved oxygen, and consequently the oxygen in the solution is used up, and passivity is destroyed. The addition of nickel improves resistance to corrosion in sea water and also improves mechanical properties. The addition of molybdenum expands the passivity range and improves resistance to neutral chlorides and sea water, but it has the disadvantage of decreasing corrosion resistance in strongly oxidizing environments.

The corrosion behavior of stainless steels in pure water up to 500 to 600°F is of importance in reactor engineering. The austenitic grades are excellent at least to 200°F, while the martensitic and ferritic grades do tarnish somewhat at these temperatures. The austenitic grades have been used for many commercial applications up to about 600°F, with good service life. They form a corrosion film at this temperature which turns into a powdery coating of red rust if there is a large amount of oxygen dissolved in the water. The use of "inhibitors" to reduce corrosion in water is frequently successful in commercial installations and may extend the service life of stainless steels in high-temperature reactor water.

The service life of stainless steels in steam does not seem to be affected by pressures at least up to 1,000 psi, but the steam temperature is important. Above a critical temperature, corrosion increases rapidly. This temperature is 1670°F for 18-8 stainless but is higher for alloys with higher chromium contents. For example, the 25 chromium–20 nickel alloys can be used at higher temperatures than the 18-8 grade.

The resistance of stainless steels to (sulfur-free) oxidizing gases is largely determined by the chromium content. Once again we find a critical temperature. A steel containing 15 per cent chromium is resistant up to about 1650°F, and an 18 per cent chromium alloy is resistant up to about 1800°F.

Table 7·8 Compositions of some high-temperature materials

Alloy	Type	% Cr	% Co	% Ni	% W	% Mo	% C	% other*
19-9DL	A	19	......	9	1.2	0.4	0.10	0.4 Ti, 0.4 Cb
γ-Cb	A	16	......	25	...	3	0.20	3 Cb
16-25-6	A	16	......	25	...	6	0.10	0.15 N$_2$
Hastelloys B	B	..	......	66	...	28	0.10	5 Fe
Inconel W	B	14	......	75	...	....	0.10	2.5 Ti, 0.6 Al
N155	B	20	20	20	2	3	0.15	1 Cb
N155	B	20	20	20	2	3	0.40	1 Cb
S590	B	20	20	20	4	4	0.40	4 Cb
S816	B	20	43	20	4	4	0.40	4 Cb
Vitallium	C	28	Balance	2	...	6	0.25	
Vitallium	C	28	Balance	2	...	6	1.00	
N155	C	20	20	20	2	3	1.00	1 Cb
422-19	C	28	Balance	12	7.5	....	0.40	

* Unless otherwise specified, the balance is iron.

The stainless steels are not generally desirable for containing liquid metals, but an important exception is their good resistance to liquid sodium and NaK. The ferritic, 18-8 austenitic, and 25 chromium–20 nickel (type 310) grades have good resistance to NaK up to 1650°F if the oxygen content of the liquid NaK is below about 0.02 per cent.

The corrosion resistance of the many grades of stainless steels in a large variety of media is known. Detailed information for a particular application can usually be obtained from one of the major producers of stainless steel, such as the Allegheny-Ludlum Steel Corporation.

PART 4. HIGH-TEMPERATURE MATERIALS

7·25 Composition

From the standpoint of neutron economy, most high-temperature materials will find use mainly in intermediate and fast reactors. Titanium, vanadium, Inconel, and the Hastelloys are included in this group. Table 7·8 gives the approximate compositions of a few of the alloys used for high-temperature service.

If the compositions of some of these alloys appear random, it is because of the tremendous difficulties inherent in the investigation of multicomponent alloys. For this reason, the development of "super alloys" is still partly an art.

7.26 Properties

The 19-9DL alloy is used for turbine wheels and contains 19 per cent chromium and 9 per cent nickel. The 16-25-6 contains 16 per cent chromium for oxidation resistance, 25 per cent nickel to make it fully austenitic, 6 per cent molybdenum for high-temperature strength, and 1 per cent silicon for oxidation resistance. The A group of alloys are designed for operation at about 1300°F. The heat-treatable group B alloys include S816, N155, and Inconel. The group C alloys are generally based on cobalt and are ordinarily used in the form of precision castings, since mechanical working is very difficult.

Table 7.9 Stress-rupture data for some typical alloys

Alloy	Condition	Stress to cause rupture in 1,000 hr, psi		
		1200°F (650°C)	1500°F (815°C)	1700°F (930°C)
18-8	Wrought, quenched, aged	11,500	3,500	
18-8 Mo	Wrought, quenched, aged	25,000	7,000	
19-9DL	Wrought, quenched, aged	37,000	10,000	
γ-Cb	Wrought, quenched, aged	36,000	11,000	
16-25-6	Wrought, quenched, aged	38,000	10,000	
Inconel W	Wrought, quenched, aged	38,000		
N-155	(0.15C) quenched, aged	38,000	13,500	
N-155	(0.406) quenched, aged	......	16,500	5,000
S-590	Quenched, aged	42,000	15,000	5,500
S-816	Quenched, aged	......	18,000	6,000
Vitallium	(0.2C) cast	44,000	16,000	10,000
Vitallium	(1.0C) cast	......	24,000	13,000
422-19	Cast	......	22,000	11,500

The mechanical properties of many of these alloys are extremely sensitive to hot working and heat-treatment. Many of them must be solution-treated, quenched, and aged to obtain optimum properties. Others, such as 19-9DL, are sensitive to small variations in composition or hot-working procedure. But the difficulties involved are not excessive if the end result is an alloy which will meet the desired requirements. Vitallium, for example, has a rupture life of 1,000 hr at a stress of 10,000 psi at 1700°F! Other rupture properties of these materials may be found in Table 7·9.

7·27 A recommendation

Although the development of super alloys is still in its infancy, some generalities are possible. It has been found that fine-grained structures are inferior to coarse-grained structures for high-temperature creep applications. This is the main reason why cast structures (large grain size) are generally superior to wrought structures. It is known that stable alloys are superior to alloys which undergo phase changes in service, that austenitic alloys are superior to ferritic alloys, and that high-melting-point elements form the strongest high-temperature alloys. Likewise it has been found that the alloy strength at any one temperature or creep rate cannot be used to predict the strength at another temperature and creep rate.

This chapter on Construction Materials is best concluded with a recommendation. We have seen that there are many properties which must be considered in choosing an alloy for a reactor application. The material which is chosen is a compromise among all properties. Since a detailed knowledge of these properties is a necessity, it is recommended that the student acquaint himself with some of the literature and with the data available from metal producers.

MODERATOR AND REFLECTOR MATERIALS

8.1 Introduction

The primary requirement for a moderator is that it be made from a material of low atomic weight. All elements, from hydrogen to carbon, atomic weights 1 to 12, are potential candidates for moderating materials. However, because of high neutron absorption cross section or low atomic density, all materials must be disqualified except hydrogen, deuterium, beryllium, and graphite. In the case of thermal and intermediate reactors, these materials can also qualify as thermal neutron reflectors, as discussed in Chap. 2.

These four moderating elements, although possessing the desired combination of low atomic weight and low thermal neutron absorption cross section, must also be made and used in a form which enables some of the mechanical load of the reactor to be transmitted through them. For this reason, graphite and beryllium must be fabricated in such a form as to impart a certain degree of strength to the moderator. In the case of hydrogen or deuterium which is utilized in the form of H_2O or D_2O, part of the reactor load is transmitted to the vessel containing water.

In addition to possessing the required cross-section values themselves, moderators must be virtually free of any impurities which would, in themselves, absorb large quantities of neutrons and which would hence reduce the efficiency of the chain reaction. Accordingly, moderating materials must be produced with special care, extra steps often being added to the processing of water, heavy water, beryllium, and graphite to remove all but minute traces of high-absorption impurities. Because moderators and reflectors play such an important role in nuclear power reactors, brief discussions of the properties of and fabrication techniques for graphite and beryllium will be presented. A discussion of water and heavy water would involve a rather detailed treatment of basic chemistry problems and is therefore not given.

PART I. GRAPHITE

8·2 Types and general properties

Graphite plays an important role in nuclear-reactor technology as a moderator and a reflector because of its low atomic weight, low neutron absorption cross section, high neutron scattering cross section, good thermal conductivity, strength and creep resistance at elevated temperatures, and high thermal shock resistance. Manufactured carbon products are predominantly of two types (not considering the artificial diamonds made by GE), both of which may be made of the same raw materials. When the material is heated only to 2750°F, it is known as *industrial carbon*. It is extremely hard and has a low thermal conductivity and a high electrical resistivity. When the material is heated to approximately 5000°F, it is known as *artificial graphite, electrographite*, or simply *graphite*. This product is easily machined and has a high thermal conductivity and a low electrical resistivity. Large quantities of both types of carbon are used commercially. However, owing to the more desirable properties and the somewhat greater purity of artificial graphite, the data in this chapter pertain primarily to this form.

The excellent high-temperature properties of graphite make the material of potentially great value in high-temperature reactors and possibly in auxiliary associated equipment (heat exchangers, etc.). In the first nuclear reactor built at the University of Chicago, graphite was used as a moderator. At present, it is used in the Hanford reactors, where each channel holding uranium fuel slugs is completely encased in a graphite block. Among the problems faced in using graphite are those due to radiation damage at low temperatures and corrosion by certain fluids at high temperatures.

8·3 Abundance and preparation

In abundance carbon is estimated to rank twelfth or thirteenth among the elements. It occurs naturally in the free state as diamond and natural graphite and is found in combination with other elements as petroleum, coal, natural gas, carbon dioxide, vegetable matter, and animal matter. Natural graphite exhibits a crystalline structure, the crystals varying greatly in size and arrangement, and contains a number of impurities which make it unusable as a reactor material.

Commercial graphite is prepared by shaping a mixture of carbonaceous filler (petroleum coke) and a binder (coal-tar pitch). The mixture is formed to shape by molding or extrusion and is then baked in gas or electric furnaces to carbonize the body completely at temperatures between 1450 and 2750°F for a period of 5 to 10 days. It is then slowly cooled for a period of time as long as 20 days or more. At this stage, the product is known as industrial carbon and is used in some applications in this form. The carbonizing treatment is followed by a reheating operation in an electric furnace at approximately 5000°F for another 3 to 6 days; it is then cooled slowly, which may take 30 to 60 days, and during the high temperature portion of the heating cycle *graphitization* takes place. During this operation, the extremely small graphite crystallites grow at the expense of the intercrystalline carbon and develop a uniform structure.

The purity of the product depends on the purity of the raw materials and the temperature of the heat-treatment. Commercial carbon products (those heated to 2,700°F or lower) vary in ash content, after complete combustion, from 0.5 to 9 per cent. Artificial graphites vary from 0.04 to 0.5 per cent. The residual contaminating impurities, listed in Table 8·1, usually are present

Table 8·1　Ash and impurity contents of commercial graphite

Grade	Ash, ppm	Chemical analysis, ppm					
		B	Al	Cr	Ti	Fe	V
Commercial graphite:							
AGX	1,500	1.3	20	2,000	50	40	70
C-18	1,200–2,000	1.0	10	1,200	35	100	200
Normal reactor-grade graphite:							
AGHT	700–1,700	...	..	500	100	100	100
AGOT	350–700	...	..	100–200	100	100	100
CS	350–700	...	..	....	...	...	100

as the carbide or hydroxide, depending on the stability of the carbide in moist air.

Reactor-grade graphite is fabricated in a manner similar to commercial-grade graphite, with minor modifications in the processing to eliminate impurities. Contaminants, which may act as neutron absorbers, are eliminated by careful selection of the raw materials. Uniformity of structures and properties are absolute "musts" for the application of graphite as a moderating material.

8·4　Physical properties

Table 8·2 shows some of the physical properties of reactor-grade graphite. The graphite crystal structure consists of layer planes of hexagonally arranged atoms, with tight binding in the planes and loose binding between the planes. When graphite is heated above 6600°F, it changes from a solid to a gaseous phase. This phenomenon is known as *sublimation*. Graphite does not form a liquid phase at any temperature at 1 atm (14.7 psi) but may form a liquid phase under pressures greater than 100 atmospheres.

8·5 Mechanical properties

The tensile and compressive strength of carbon and artificial graphite are given in Table 8·3. The values quoted cover a wide range. When the design

Table 8·2 Physical properties of reactor-grade graphite *

Density, gm/cm^3:
Calculated from lattice constants 2.27
AGHT 1.61–1.65
AGOT 1.62–1.72
Melting point, boiling point . . . . .Sublimes
Sublimation temperature . . . . .6600 $\pm$ 50°F, 3650 $\pm$ 25°C

Electrical resistivity, microhm-cm:
AGHT 800
AGOT 500–850
CS 800
Coefficient of thermal expansion, per °C at 20°C . . . .1.5–7.5 $\times$ 10^{-6}
Thermal conductivity, cal/sec-cm-°C 0.35

Crystal structure (many different forms):
Hexagonal (generally accepted), Å . . . .$a_0 = 2.456$
$c_0 = 6.697$

* Thermal neutron absorption cross section, AGOT, barns per atom 4.8 $\times$ 10^{-3}.

engineer requires a specific value, the problem should be reviewed with the manufacturer of the carbon or graphite and a product selected for the individual application. The strength of artificial graphite varies with

Table 8·3 Mechanical properties of carbon and artificial graphite at room temperature

Type	Tensile strength, psi		Compressive strength, psi	
	Longitudinal	Transverse	Longitudinal	Transverse
Commercial carbon	1,000	500–3,000	2,000–10,000	2,000–8,000
Commercial graphite . . .	500–2,000	500–3,000	. . .	3,000–6,500
AGHT	2,200	600		

the anisotropic nature of the material and with its bulk density. During the fabrication of graphite shapes, the basal planes orient in the direction of the extrusion forces or, in molding, perpendicular to the molding forces. The strength of graphite also depends on this orientation. It is considerably stronger in the direction of preferred orientation than in any other direction.

Graphite is unique among materials in exhibiting much greater strength at elevated temperatures than at room temperature. This is shown for the tensile strength in Fig. 8·1. The compressive strength responds similarly with increasing temperature. There are few materials which are as strong as graphite at temperatures 5000°F or above, a fact which makes graphite extremely interesting as a high-temperature material. The thermal shock resistance of graphite is outstanding; it can hardly be fractured by thermal stresses.

8·6   Corrosion behavior

Because of its use as a crucible material, the corrosion behavior of graphite in contact with liquids has been extensively investigated and is well known. A partial list showing the reactions of graphite with several gases, aqueous solutions, and solids is given in Table 8·4.

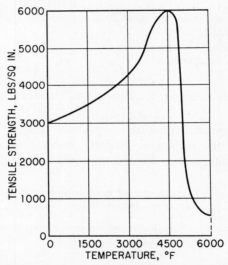

Fig. 8·1   Schematic representation of the effect of temperature on the tensile strength of graphite.

PART 2.   BERYLLIUM

8·7   General properties

The metal beryllium differs greatly from most other metals. It is the only light metal (density = 1.82 gm/cm³) with a high melting temperature (1315°C). It is extremely penetrable to X rays and is an excellent transmitter of sound, having a sound-velocity value twice that of aluminum or steel. From a reactor-application viewpoint, beryllium makes an excellent moderator or reflector because of its low thermal neutron absorption cross section and high neutron scatter cross section. Furthermore, the neutron multiplication may be increased by a small increment through beryllium (γ,n) and $(n,2n)$ reactions. However, these useful properties of beryllium are somewhat balanced by its low ductility, difficulty of fabrication, generally poor corrosion resistance to water and the various cooling media of proposed reactor designs, and high cost.

Table 8·4   Reaction of graphite with corrosive agents

Agent	Temperature, °F	Reaction Products
Gases:		
Oxygen	840 and higher	Produces carbon oxides
Hydrogen	1650–1830	Methane
Water Vapor	1470 and lower	Negligible reaction
	1470 and higher	Increasingly severe attack
Aqueous solutions:		
Dilute acid or alkali	Less than boiling	No attack
Strong acid (HNO_3 and H_2SO_4)	570 and higher	Graphitic oxide
KOH (50% solution)	660 and higher	Dissolves graphite
Solids:		
Metals (most).............	2730 and higher	Metal carbide
Metal oxides	2730 and higher	Metal carbide and carbon oxides

Occasionally, a sample of beryllium will possess fairly good ductility and show good corrosion resistance in water, indicating possibly that some of the poorer properties ascribed to beryllium might result from impurities in the metal. No crystallographic basis for beryllium's lack of ductility appears to exist, although its small atomic volume would lead one to expect that it would be harder than other hexagonal metals.

8·8   Abundance and preparation

The mineral *beryl* (a beryllium aluminum silicate, $3BeO–Al_2O_3–2SiO_2$), the only commercial source of beryllium, is recovered primarily by hand sorting, although recently developed concentration processes show promise. The mineral is not found in large deposits. Its occurrence is erratic and difficult to predict. The principal producers of beryl ore are the Union of South Africa, Southern Rhodesia, Brazil, India, and the United States.

Although the element beryllium was discovered as early as 1798, no sizable beryllium industry was established before 1930. Practically all pure beryllium metal now produced in the United States is made by the reduction of beryllium fluoride with magnesium metal. The beryllium fluoride used is produced from beryllium oxide. In this process, developed by Kjellgren and used by the Brush Beryllium Company, the final product is in the form of spheroids known as "beads," or "pebbles," which are produced in high-purity as well as commercial grade. Another process for making high-purity beryllium consists in melting beryllium chloride with sodium chloride and forming a fused electrolyte, from which beryllium can be prepared by electrolysis in the form of "flakes." The pebble beryllium is extremely brittle, whereas the flake beryllium usually possesses some ductility.

8·9 Production of beryllium metal

Beryllium-metal slabs or billets are made by vacuum melting and casting or by powder-metallurgy techniques. It is customary to melt beryllium in vacuum and to use beryllia (BeO) crucibles. The vacuum is required, not merely to minimize oxidation, but also to remove volatile constituents such as magnesium, sodium, and various fluorides and chlorides. Beryllia crucibles

Table 8·5 Some room-temperature directional properties of extruded, cast, and flake beryllium

Material	Condition	Longitudinal to extrusion		Transverse to extrusion	
		Ultimate strength, psi	Elongation, per cent	Ultimate strength, psi	Elongation, per cent
Cast	As extruded	32,700	0.36	19,400	0.3
Extruded	Annealed 800°C	40,000	1.82	16,600	0.18
Extruded	As extruded	46,600	0.55	29,100	0.3
Flake	Annealed 800°C	63,700	5.0	25,500	0.3

are required because the metal will react to a certain extent with all other stable oxides, especially in vacuum. Degassed graphite is the usual mold material for beryllium castings.

Beryllium castings have been used mainly for producing extrusion billets. The metal as cast is exceedingly brittle, largely because of the coarse grain size. A casting of intricate shape would almost certainly contain internal cracks at the regions where the *columnar grains* meet (large grains which grow from the surface of the mold in toward the center of the casting). The machining characteristics of castings are poor because the hardness of individual crystals is quite anisotropic and in addition surface cracks form readily. These difficulties could be avoided if the casting could be made sufficiently fine-grained. There is considerable incentive to achieve fine-grained castings, since the cast metal is much less expensive than the powder-metallurgy product.

The powder-metallurgy product made by the Brush Beryllium Company, known as QMV, is now the principal form of beryllium suitable for commerical fabrication. This material is made by vacuum hot pressing at 1000 to 1150°C (1830 to 2100°F) and can be fabricated by extruding, rolling, or forging in the range of 1000 to 1150°C when protected by a steel sheath.

Table 8·6 Physical properties of beryllium

Thermal neutron absorption cross section,
 barns/atom. . . . . .9 × 10⁻³
Density, gm/cm³ 1.85
Melting point. . . . . .2341 ± 6°F, 1283 ± 3°C
Boiling point . . . . .5380°F, 2970°C
Electrical resistivity, microhm-cm 3.9–4.3
Linear coefficient of expansion, per °C:
 20–200°C . . . .13.3×10^{-6}
 20–7000°C . . . .17.8×10^{-6}
Thermal conductivity at 20°C, cal/sec-cm-°C. 0.835
Crystallography, hcp, Å. . . . .$a_0 = 2.2858$
$c_0 = 3.5842$
Allotropic transformation, at about . . . . .1250°C

A comparison of the mechanical properties of hot extruded, cast metal, and flake is given in Table 8·5. The material fabricated from powder has the best mechanical properties achieved for beryllium to date. Fine-grained, hot-pressed, or fabricated powder metal, in general, produces a material which can be readily machined, while the cast material can be machined only with difficulty.

8·10 Physical properties

The physical properties of beryllium are listed in Table 8·6. Thermal measurements indicate a phase transformation approximately 30° below the melting point. High-temperature X-ray measurements are in progress to determine whether or not the heat effect is due to an allotropic transformation. Some investigators believe that the cause may be local melting of a eutectic mixture of beryllium and beryllium oxide.

8·11 Mechanical properties

The mechanical properties of beryllium are quite dependent on the degree of anisotropy and the purity of the metal. Extruded beryllium is anisotropic because of the preferred orientation introduced. This orientation persists to some extent even after annealing at 1830°F. However, no evidence of anisotropy is observed in the mechanical properties which have been produced by hot pressing powder.

Ordinarily, beryllium is a metal of fairly high tensile strength, high modulus of elasticity (higher than that of steel), and low ductility. There appears to be no fundamental crystallographic reason for the poor ductility. Its similarity in structure to titanium and zirconium leads one to expect beryllium to be ductile in the pure state. Despite considerable effort to increase its purity, the unalloyed metal remains quite brittle.

Table 8·5 presents the room-temperature tensile data for various types of beryllium. The data illustrate the marked differences between transverse and longitudinal properties in extruded, cast billets at temperatures above 450°C is considerably greater than that of extruded flakes.

8·12   Joining

The problems associated with the brazing of beryllium are, in many respects, similar to those encountered in the brazing of aluminum. Because of its high affinity for oxygen, beryllium requires careful preparation and handling during the brazing operation. Beryllium has been furnace-brazed successfully in an argon atmosphere at temperatures between 1400 and 1600°F, foils of aluminum or of one of several aluminum alloys being used. Beryllium has been soldered in the normal manner after it has been coated with copper. The copper may be applied by painting the beryllium sample with a paste of $CuCl_2$ in turpentine. When this mixture is heated to 800°F, a copper deposit is left on the beryllium.

Attempts to weld beryllium by means of a hydrogen arc have been unsuccessful. Helium arc welding in an argon arc, however, can be accomplished satisfactorily.

8·13   Corrosion behavior

Beryllium does not corrode appreciably in air below 750°F. At higher temperatures (about 1500°F), a white film forms, presumably BeO, which does not flake off. According to some reports, the resistance of beryllium to oxidation at 1290°F appears to equal that of 18-8 stainless steel. The oxidation resistance of extruded vacuum-cast beryllium is reported to be superior to either the extruded flake or the powder-metallurgy (QMV) product at 1290 to 1470°F in air.

Beryllium appears to be definitely inferior to zirconium in corrosion resistance to water. The corrosion rate is increased when the pH is raised above 6.5 or if the extruded surface has been chemically etched prior to exposure. The presence of copper or chlorine in aqueous solution is known to cause deep pitting in beryllium. Aluminum ions do not appear to have any adverse effect, even at appreciable concentrations. Chemical inhibitors have been found which decrease corrosion of beryllium in reactor water. Beryllium also shows poor corrosion resistance in liquid metals when they are open to air. In oxygen-free liquid-metal systems, beryllium is much more corrosion-resistant.

BIBLIOGRAPHY

Barrett, C. S.: "Structure of Metals," McGraw-Hill Book Company, Inc., New York, 1952.

Brick, R. M., and A. Phillips: "Structure and Properties of Alloys," McGraw-Hill Book Company, Inc., New York, 1949.

Davies, A. C.: "The Science and Practice of Welding," University Press, Cambridge, Mass., 1956.

Doan, G. E.: "The Principles of Physical Metallurgy," McGraw-Hill Book Company, Inc., New York, 1953.

Hampel, C. A.: "Rare Metals Handbook," Reinhold Publishing Corporation, New York, 1961.

Kaufmann, A. R.: "Nuclear Reactor Fuel Elements," Interscience Publishers, New York, 1962.

"Liquid Metals Handbook," U.S. Atomic Energy Commission, 1952.

Lustman, B., and F. Kerze, Jr.: "The Metallurgy of Zirconium," McGraw-Hill Book Company, Inc., New York, 1955.

"Metals Handbook," American Society for Metals, Cleveland, Ohio, 1948.

Schuhmann, R., Jr.: "Metallurgical Engineering," vol. 1, Addison-Wesley Publishing Company, Inc., Cambridge, Mass., 1952.

Seitz, F.: "The Physics of Metals," McGraw-Hill Book Company, Inc., New York, 1943.

Slater, J. C.: "Quantum Theory of Matter," McGraw-Hill Book Company, Inc., New York, 1960.

Taggart, A. F.: "Handbook of Mineral Dressing," John Wiley & Sons, Inc., New York, 1945.

Uhlig, H. H.: "The Corrosion Handbook," John Wiley & Sons, Inc., New York, 1948.

ANSWERS TO PROBLEMS

Chapter 3

1. *d* 2. *b, c* 3. *b, c* 4. *b* 5. *a, c, d* 6. *d* 7. *a, c, d* 8. *a, b, d*
9. *b, c* 10. *a, c* 11. *b, d* 12. *a, d*

Chapter 4

1. *a, b, d* 2. *a, b* 3. *a, b, d* 4. *a* 5. *b, d* 6. *a, b, c* 7. *c* 8. *a, c*
9. *a, c, d* 10. *b, d* 11. *c, d* 12. *c* 13. *a, c* 14. *a, b, d* 15. *b, c, d*
16. *b, c* 17. *c, d* 18. *c* 19. *a, d* 20. *a* 21. (*a*) 183°C, 11.9%, eutectic;
room temperature 11.9%, eutectic. (*b*) 183°C, α is 80% Pb, liquid is 62% Sn;
room temperature, α is 99% Pb, β is 99% Sn 22. (*a*) One; (*b*) two;
(*c*) *y*, β + δ, α + δ 23. (*a*) Completely miscible in liquid and solid state.
(*b*) 1400°C, liquid 50% Cu (100%); 1300°C, liquid 55% Cu (78.3%); α,
32% Cu (21.7%); 1200°C, α 50% Cu (100%) 24. (*a*) Eutectic, none;
eutectoid, one; peritectic, five; peritectoid, none. (*b*) 900°C, liquid; 800°C, β;
500°C, β + γ; 400°C, β′ + γ; (*c*) Liquid, 29.3%; γ, 60.7%.

Chapter 5

1. *b, c* 2. *a, c* 3. *a, b*, possibly *c* 4. *b, c, d* 5. *c* 6. *a* 7. *b, c, d*
8. *a, c, d* 9. *b, c* 10. *a, b* 11. *a, d* 12. *c, d*

INDEX